A PRIMER OF GENETICS

A. RADFORD, D. J. COVE & S. BAUMBERG
DEPARTMENT OF GENETICS
UNIVERSITY OF LEEDS

Longman
Longman Group Limited
Longman House, Burnt Mill, Harlow,
Essex CM20 2JE, England
and Associated Companies throughout the world

First published 1995

British Library Cataloguing in Publication Data
A catalogue entry for this title is available from the British Library.

ISBN 0-582-21718-0

Library of Congress Cataloging-in-Publication Data

A catalog entry for this title is available from the Library of Congress.

Set by 6 in 10 on 12pt Bembo
Produced by Longman Singapore Publishers (Pte) Ltd.
Printed in Singapore

CONTENTS

Contents

Chapter 7   Advanced Linkage Analysis

PREFACE

Genetics is generally perceived to be a difficult subject. It is true that it requires a logical and methodical approach, probably to a greater extent than other branches of biology. However, once that approach is mastered, the subject benefits from the lack of any great factual basis to be committed to memory. Once the principles are understood, it is possible to derive the detail.

The perception that genetics is difficult derives from the attempt by some to teach or learn the subject by rote. This approach is doomed to failure. A superficial familiarity with the subject, with the vocabulary, does not lead to understanding. The acid test comes with the ability to apply genetic principles and solve problems.

As with any problem of data interpretation, it is necessary to learn the way to approach the data. Invariably, a large mass of data are presented, far too much to analyse all at once. The skill is in identifying the smallest meaningful datum, and then adding additional data progressively to permit the development of the model or the elimination of alternatives.

To facilitate the development of the necessary genetic problem-solving skills, this book presents a series of solved problems. By carefully following the method, it is hoped that the student will acquire the necessary appreciation to solve problems on all aspects of the subject.

Of course, in a book of this size, it is not possible to illustrate all possible types of questions and their solutions. However, the mastery of genetic problem-solving, and the broader field of the interpretation of data in genetic research, have a commonality of approach and method.

The questions in this book or others very similar to them have been used for a number of years by the authors and their colleagues in the teaching of undergraduate students in the Department of Genetics of the University of Leeds.

1 MENDELIAN GENETICS

Introduction

Although, in the middle of the last century, Gregor Mendel solved the problems of segregation and independent assortment as aspects of theoretical probability, it is certainly not necessary to do so today. Segregation, independent assortment and also linkage are consequences of the process of meiosis, about which Mendel was ignorant. Therefore, we can take advantage of the understanding of meiosis and associate these genetic processes with their cytological basis.

Meiosis

Any genetic textbook has a description, and diagrams, of the normal eukaryotic life cycle, including the process of meiosis (Fig. 0.1). These will illustrate the way in which two normally haploid gametes, one from each parent, fuse to form the diploid

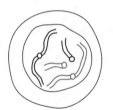

1. Leptotene

2. Zygotene

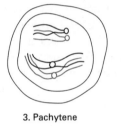

3. Pachytene

4. Diplotene

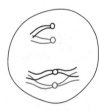

5. Diakinesis

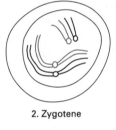

6. Metaphase I

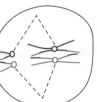

7. Anaphase I

8. Telophase I

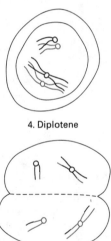

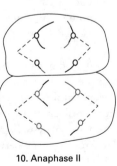

9. Metaphase II

10. Anaphase II

11. Telophase II

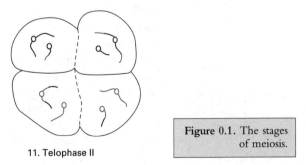

Figure 0.1. The stages of meiosis.

zygote. The zygote, and any cells mitotically derived from it, have identical chromosome content – two copies of each chromosome (except perhaps the sex chromosomes, if present), one from each parent. These two copies are homologous, which means that they are similar but not necessarily genetically identical. Differences between them may have arisen by mutation and other sources of genetic variation.

In meiosis, each chromosome replicates, its DNA genetic content replicating semi-conservatively, to give rise to two identical copies (sister chromatids). Homologous chromosomes (each now two sister chromatids) associate in the early stages of meiosis, aligning all four chromatids. Breakage and rejoining may give cytologically visible chiasmata and their genetic equivalent, crossovers, at this stage (pachytene/diplotene of prophase I).

Two successive cycles of division reduce the temporarily tetraploid nuclear content to four daughter cells, each haploid. In the first division, homologous centromeres move to opposite poles. At the second division, the centromeres themselves divide and pull pairs of sister chromatids (subject to any non-identical regions due to crossing over) apart. In this way, for any gene heterozygous in the initial diploid cell (i.e. because its two parental gametes contained different alleles of the gene in question), the result is normally two haploid cells of each allele type in the four meiotic products. This is the basis of segregation, described by Mendel's first law. There are exceptions to this rule (normally rare) that tell us about the precise mechanism of the recombination process, but these will be considered later.

For any given synapsed pair of chromosomes (four chromatids) in early meiosis (prophase I), their orientation with respect to the poles of the forthcoming division when entering metaphase I is random. Thus when two non-homologous chromosomes are each heterozygous, all relative orientations on the spindle are equally probable. This is the basis of independent assortment, described by Mendel's second law.

Genetic analysis   As mentioned above, crossing over between non-sister chromatids in meiotic prophase I gives genetic exchange. This leads to segregation of a single heterozygous gene, and to assortment (but not independent) between the alleles of each of two heterozygous genes on the same chromosome. Because the frequency of occurrence of crossing over in an interval is related (but not necessarily proportional) to the physical length of DNA double helix between the two sites, the frequency of crossing over in an interval determines its genetic length. In this way, the distance between a gene and the centromere of the chromosome on which it is located can be measured. Likewise, the distance between two genes on the same chromosome can be determined.

In certain organisms, especially certain fungi such as *Sordaria*, *Neurospora* and the yeast *Saccharomyces*, it is possible to isolate all four products of individual meiotic events and analyse them. This permits the special form of genetic analysis known as tetrad analysis, which gives additional information about genes and centromeres. In the first two genera, it is possible to obtain from the analysis, because of the linear nature of the tetrads, data directly attributable to crossing over between any single gene and its centromere (Fig. 0.2).

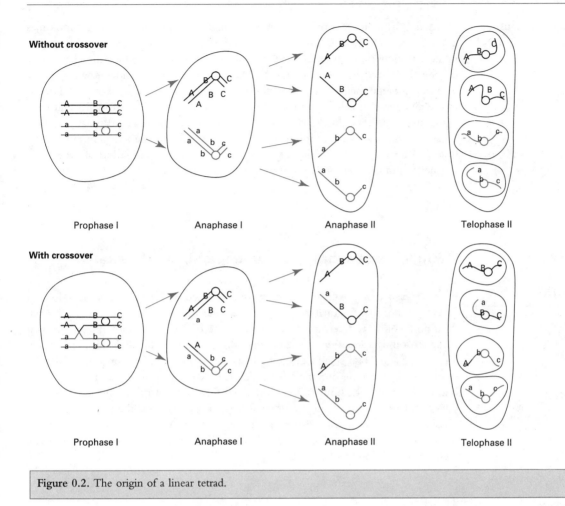

Figure 0.2. The origin of a linear tetrad.

As sister chromatids are both attached to the same centromere, they should stay together, but segregate from the other pair of sister chromatids, at the first division (anaphase I) of meiosis. The only exception arises if there has been a crossover between the gene and its centromere, when a second division segregation will result. There is the possible complication of multiple crossovers in an interval, which might lead to a crossover not being detected, but this will be considered later.

The problems in this first chapter will cover aspects of segregation, independent assortment and linkage. They will exemplify data from linear tetrads, unordered tetrads and random meiotic products. They will further include examples of data from organisms in which haploid meiotic products may be directly analysed, and others in which scoring must be done on diploids. For analysis of such diploids, additional complications of dominance, plus the necessity to test-cross or back-cross, are involved.

Non-chromosomal DNA

Of course, the major part of the genetic information is contained in the chromosomes in the nucleus of the cell. However, other information-carrying DNA is also present in the cell and is transmitted from cell to cell and generation to generation. This DNA is generally found in cytoplasmic organelles, the chloroplasts and mitochondria. It does not undergo meiosis, but replicates with the organelles and segregates with them in the cytoplasm at cell division. In the production of gametes, organelles and their DNA may again be included with the chromosomal genetic material. In most organisms, the male gamete is much smaller than the female gamete. Thus in gamete fusion to form the zygote, the major and possibly only contribution of cytoplasm, organelles and organelle DNA is from the female parent.

Problem 1

Single gene segregation in haploid, linear tetrads

Question

In *Neurospora crassa*, wild-type meiotic spores (ascospores) are black in colour; a mutant strain that is unable to synthesize melanin produces white ascospores. Ascospore colour is determined by the genotype of the nucleus it contains. A slight complication is caused by the fact that there is a mitotic division immediately following the meiosis, so an ascus (tetrad) actually contains eight spores (each adjacent spore pair representing one meiotic product). A *Neurospora* ascus is a linear tetrad, so one half of the ascus (four spores) is derived from one of the products of the first division of the meiosis, and the other half (four spores) from the other product of division 1 of meiosis. A wild-type strain is crossed to a white-spored (*asco*) mutant strain. Of 100 asci (tetrads) scored by microscopic observation, the following types of segregation are observed:

i	ii	iii	iv	v	vi
44 asci	36 asci	5 asci	7 asci	4 asci	4 asci
asco	+	*asco*	*asco*	+	+
asco	+	*asco*	*asco*	+	+
asco	+	+	+	*asco*	*asco*
asco	+	+	+	*asco*	*asco*
+	*asco*	*asco*	+	*asco*	+
+	*asco*	*asco*	+	*asco*	+
+	*asco*	+	*asco*	+	*asco*
+	*asco*	+	*asco*	+	*asco*

Interpret these data as fully as possible.

Answer

The first observation is that all asci contain four wild-type (+) spores and four of the mutant (*asco*) type. Therefore, in all cases meiosis has given rise to two meiotic products of each genotype. This both confirms Mendel's first law and that we are indeed studying the results of segregation of a single heterozygous gene.

Furthermore, because these are linear tetrads, if we classify them into first-division segregations and second-division segregations, we can determine the genetic distance of the *asco* gene from its centromere.

There are 44 + 36 (80) first-division segregation tetrads (classes i and ii), all the rest (5 + 7 + 4 + 4 = 20) being second-division segregations (classes iii, iv, v and vi).

In order to obtain a second-division segregation, it is necessary to have a crossover between two non-sister chromatids in the interval between the gene and its centromere. It is important to note that in such a second-division segregation tetrad, only two of the four chromatids are crossed over.

Of the 100 tetrads scored, 20 must contain a crossover. In each of those 20, half of the chromatids are crossed over and half are not. Therefore the frequency of crossing over in the interval between *asco* and its centromere is a half of the frequency of second-division segregation tetrads. Recombination frequency is conventionally expressed as a percentage (also called map units). Therefore the distance between centromere and *asco* is:

$$\frac{1}{2} \times \frac{20}{100} \times 100\%$$

$$= 10\% \text{ recombination}$$

$$= 10 \text{ map units}$$

From these data we have determined that the *asco* (white ascospore) phenotype segregates in a 1:1 ratio with wild type (black ascospore), consistent with the two phenotypes being determined by two alleles of a single gene locus. We have further determined the map position of the responsible gene locus with respect to the centromere of the chromosome on which it is located.

| Problem 2 | **Single gene segregation in haploid, random progeny** |

Question

In the fungus *Aspergillus nidulans*, the haploid meiotic products are analysed directly. However, because of technical difficulties, few scientists have managed to carry out tetrad analysis and crosses are generally analysed from samples of random meiotic progeny. In a cross between two strains, one wild type and the other with a requirement for the vitamin biotin for growth, analysis of 100 random meiotic progeny gave 46 wild-type progeny and 54 that were biotin auxotrophs. Interpret these data as fully as possible.

Answer

Our first hypothesis is that the biotin requirement of one parent is the result of a single mutant blocking the biotin biosynthetic pathway, probably by failing to produce an active enzyme required to catalyse one of the interconversions in the pathway. If this is the case, we are postulating that the wild-type strain possesses an active allele and the auxotroph an inactive allele. There are thus two alleles present in the zygote (i.e. the

Degrees of freedom (n)	Probability (P)					
	0.9	0.5	0.1	0.05	0.01	0.005
1	0.02	0.46	2.71	3.84	5.02	7.88
2	0.21	1.39	4.61	5.99	9.21	10.60
3	0.58	2.37	6.25	7.82	11.35	12.84
4	1.06	3.36	7.78	9.49	13.28	14.86
5	1.61	4.35	9.24	11.07	15.09	16.80
6	2.20	5.35	10.65	12.59	16.81	18.55
7	2.83	6.35	12.02	14.07	18.48	20.28
8	3.49	7.34	13.36	15.51	20.09	21.96
9	4.17	8.34	14.68	16.92	21.67	23.59
10	4.87	9.34	15.99	18.31	23.21	25.19
11	5.58	10.34	17.28	19.68	24.73	26.76
12	6.30	11.34	18.55	21.03	26.22	28.30

heterozygous state), and in meiosis they should segregate $2:2$ in the four products of an individual meiosis.

Whereas in tetrad analysis the consequence of a $2:2$ segregation is an exact numerical $2:2$ tetrad, in random progeny we obtain a random sample that will reflect the overall ratio of $1:1$, i.e. 50% wild type and 50% biotin dependent. What we actually measured, in a sample of 100 progeny, is 46 and 54. Is this acceptable as a $1:1$ ratio? To test this hypothesis, we require a statistical test, the appropriate one being the χ^2 test. In this we estimate the probability of obtaining the observed or a worse fit to the expected ratio by sampling from a total population with the exact ratio. For reasons beyond the scope of this book, a probability of less than 0.05 (5%, or 1 in 20) is regarded as evidence that the population sampled does not have the expected composition, and therefore that our hypothesis is incorrect.

The χ^2 test compares, for each class, the difference between observed (O) and expected (E), squares the difference, and divides the difference by the expected class size. These values for each class are summed to give the value of χ^2. For the appropriate number of degrees of freedom (for a χ^2 test, one less than the number of classes), the probability corresponding to the value of χ^2 is read from the matrix of a table of values of χ^2 (Table 2.1).

The equation for the value of χ^2 is:

$$\chi^2 = \sum \frac{(O - E)^2}{E}$$

In our results, from 100 random progeny, we would expect 50 wild type and 50 biotin dependent, so this is the expected value for each class. Substituting in the above equation:

$$\chi^2 = \frac{(46 - 50)^2 + (54 - 50)^2}{50}$$

$$= \frac{16 + 16}{50}$$

$$= 0.64$$

With two classes, there is one degree of freedom. From the χ^2 table, a value of 0.64 with one degree of freedom gives a probability greater than 0.1 but less than 0.5. Our result has a probability considerably higher than 0.05, the threshold of significance. It is thus statistically not significantly different from the expected $1:1$ ratio.

Problem 3	**Single gene segregation in diploid, no dominance**

Question

In a cross between a pure-breeding red-flowered *Antirrhinum* and a pure-breeding white-flowered *Antirrhinum*, all of the F_1 progeny had pink flowers. (Note that in Mendelian pedigrees, P designates the parental generation, F_1 the next (first filial) generation and F_2 the second filial generation. Note also that, for example, the seeds produced on the (P) parent are the first stage of the F_1, and certain phenotypic characters such as seed shape and colour may be scored directly at this stage. The plant into which the seed germinates is also the F_1 and other phenotypic characters such as plant height, leaf shape and flower colour can only be scored at this late stage.)

1. Some of the F_1 plants were bagged and therefore were self-fertilized. Progeny (F_2) were harvested and grown, and the flower colours on these F_2 plants were scored. Of 100 F_2 plants, 20 had red flowers, 53 had pink flowers and 27 had white flowers.
2. Some of the F_1 plants had their anthers removed before pollen release and were subsequently fertilized with pollen from the red-flowered parental strain. The progeny of this back-cross was analysed: of 100 plants classified, 43 were red-flowered and 57 were pink-flowered.
3. Some of the F_1 plants had their anthers removed before pollen release and were subsequently fertilized with pollen from the white-flowered parental strain. The progeny of this back-cross was analysed: of 100 plants classified, 54 were pink-flowered and 46 were white-flowered.

Interpret these results as fully as possible.

Answer

This species is diploid (at the stage when phenotype is determined) so Mendelian segregation ratios cannot be determined directly from meiotic products, but only when these have combined with other gametes to restore the diploid state. Therefore, genetic analysis in diploids is inherently more difficult than it is in haploids. Hence the necessity for two generations rather than the one needed for full haploid analysis.

In the parental cross, pure-breeding red and white parents give an intermediate (pink) F_1 phenotype. In the F_2 both red and white phenotypes, as well as pink, reappear. The critical observation is the ratio of red:pink:white, which in the F_2 is $20:53:27$. This approximates the Mendelian $1:2:1$ ratio, the F_2 genotype ratio expected in diploids if a single gene is segregating. Let us test this hypothetical ratio, using the χ^2 test. In this case we have three classes, expected to be $25:50:25$, and hence two degrees of freedom. The equation is:

$$\chi^2 = \sum \frac{(O - E)^2}{E}$$

Our calculation is therefore:

$$\chi^2 = \frac{(20 - 25)^2}{25} + \frac{(53 - 50)^2}{50} + \frac{(27 - 25)^2}{25}$$

$$= \frac{25}{25} + \frac{9}{50} + \frac{4}{25}$$

$$= \frac{67}{50}$$

$$= 1.34$$

For two degrees of freedom, this value of χ^2 gives us a probability of just over 0.5, so the observed results are acceptable as the postulated $1 : 2 : 1$ ratio.

On this basis, the red-flowered plants are homozygous for one allele (p^1p^1) and the white-flowered plants are homozygous for a second allele (p^2p^2). The F_1 plants are all heterozygous (p^1p^2) because they have a p^1 allele from the gamete from one parent and a p^2 allele from the other parental gamete, and pink in colour because there is no dominance of one allele over the other.

In the F_2, the 1 red : 2 pink : 1 white ratio results from the $1 : 1$ segregation of p^1 and p^2 gametes that the F_1 plants produce and their combination in the fertilization process:

		Male gametes	
		$\frac{1}{2}\,p^1$	$\frac{1}{2}\,p^2$
Female gametes	$\frac{1}{2}p^1$	p^1p^1 (red) $\frac{1}{4}$	p^1p^2 (pink) $\frac{1}{4}$
	$\frac{1}{2}p^2$	p^2p^1 (pink) $\frac{1}{4}$	p^2p^2 (white) $\frac{1}{4}$

With no dominance, we can immediately determine the genotype of the organism from its phenotype. We can therefore identify the heterozygotes and the two homozygotes. In this case, selfing either a red-flowered plant or a white-flowered plant would show it to be true-breeding.

In the first back-cross (2) we obtained the cross red (p^1p^1) × pink (p^1p^2). A diagram of the gametes and their combination is:

		Red parent gametes
		$1\,p^1$
Pink parent gametes	$\frac{1}{2}p^1$	p^1p^1 (red) $\frac{1}{2}$
	$\frac{1}{2}p^2$	p^1p^2 (pink) $\frac{1}{2}$

Thus they give a 1 : 1 ratio of the red (p^1p^1) and the pink (p^1p^2) types. Note that there are no white progeny, as there is no way to obtain a p^2 gamete from the red parent.

In the second back-cross (3) we obtained the cross white (p^2p^2) × pink (p^1p^2). A diagram of the gametes and their combination is:

White parent gametes
$1\,p^2$

		White parent gametes $1\,p^2$
Pink parent gametes	$\frac{1}{2}p^1$	p^1p^2 (pink) $\frac{1}{2}$
	$\frac{1}{2}p^2$	p^2p^2 (white) $\frac{1}{2}$

Thus they give a 1 : 1 ratio of the pink (p^1p^2) and the white (p^2p^2) types.

For completeness, the observed back-cross ratios should be tested for their goodness of fit to the expected 1 : 1 ratio using the χ^2 test.

Problem 4	Single gene segregation in diploid, dominance

Question

In *Drosophila melanogaster*, the normal (wild-type) eye colour is dark red. A pure-breeding mutant strain with brown eyes exists. When wild-type flies are crossed to brown–eyed flies (males of one strain mated with virgin females of the other) all the F_1 progeny have dark-red eyes.

F_1 flies were inter-crossed and the resultant F_2 analysed. In a total of 200 F_2 flies classified, there were 141 wild-type flies and 59 brown-eyed flies. There was no difference in the results between male and female F_2 progeny.

Some F_1 flies were used in a test-cross to individuals of the brown-eyed strain, and of 88 progeny 48 were wild type and 40 were brown-eyed. Again, results were the same for both sexes of F_2 progeny.

Interpret these results.

Answer

Drosophila is an organism that is diploid at the stage at which it is classified. The two parental strains are true-breeding and thus presumably homozygous for eye colour. The F_1 flies are therefore presumably heterozygous for any gene by which the parents differ, which in this case is that determining eye colour. The results of the F_2, in which the brown eye phenotype appears in a minority, are consistent with this hypothesis.

Let us postulate a single gene, *b*, affecting eye colour in this case. Of the two pure-breeding parental strains, the wild type is *++* in genotype (the two *+* symbols represent the two copies of the wild-type allele of the single gene involved) and the brown-eyed strain is *b b*. The phenotypically wild-type F_1 flies should all be *+b*, because the *+* (wild-type) allele is dominant with respect to the *b* allele.

Inter-crossing F_1 flies should give the following results:

Female gametes

		$\frac{1}{2}+$	$\frac{1}{2}b$
Male gametes	$\frac{1}{2}+$	$++$ wild $\frac{1}{4}$	$+b$ wild $\frac{1}{4}$
	$\frac{1}{2}b$	$b+$ wild $\frac{1}{4}$	bb brown $\frac{1}{4}$

This predicts a genotypic ratio of $1 ++ : 2 +b : 1 bb$ and a phenotypic ratio of 3 wild type : 1 brown-eyed. The latter we can investigate directly by comparing our actual data with the result predicted from our hypothesis, using the χ^2 test.

Of the 200 F_2 flies scored, 141 were wild type and 59 were brown-eyed. If the 3 : 1 ratio is true, we would expect there to be 150 and 50 in the two classes respectively. Therefore:

$$\chi^2 = \frac{(141 - 150)^2}{150} + \frac{(59 - 50)^2}{50}$$

$$= \frac{81}{150} + \frac{81}{50}$$

$$= 2.16$$

For one degree of freedom, a χ^2 value of 2.16 has probability of greater than 0.1, so our hypothesis is acceptable.

From our hypothesis, the test-cross would result in a progeny phenotypic ratio of 1 wild type : 1 brown-eyed. The actual results were 48 and 40 in the two classes respectively. Again, a χ^2 test gives a value of about 0.75, which with one degree of freedom has a probability greater than 0.1, again therefore consistent with our hypothesis.

The problem can therefore be explained by single gene segregation, the mutant b (brown-eyed) allele being recessive to the wild type.

Problem 5 — Single gene segregation in diploid, recessive lethal

Question

In mice, there are many coat colour mutants now known. One of the early ones was a paler than normal colour, called *yellow*. One problem with mammals, compared with bacteria, fungi or *Drosophila*, is that litter sizes are too small for individual statistical analysis. Phenotypic ratios, if required, must be determined from the summed results of possibly many litters from the same or identical matings.

In all matings of *yellow* × *yellow* mice, wild-type and *yellow* progeny are produced. From the sum of five different litters from such matings, the progeny were 12 wild type and 33 *yellow*. In a series of back-crosses of F_1 mice, the following results were accumulated:

F$_1$	Back-cross parent	Back-cross progeny	
		wild type	yellow
yellow	yellow	330	620
yellow	wild type	330	380
wild type	yellow	409	450
wild type	wild type	635	0

From several litters from each of the 33 *yellow* F$_1$ mice, crossed to either *yellow* or wild type, both phenotypes were present in the progeny.

Explain these results as fully as possible.

Answer

At first sight, the initial cross appears to be between two heterozygous individuals, as wild-type progeny segregate from two *yellow* parents. The *yellow* allele may be designated Y, capitalized to indicate its dominance over the wild-type allele $+$. Note that there is no necessity for 'wild type' to be dominant over mutant alleles, although it is in most cases. If *yellow* is dominant over wild type in this case the two parents are both of $Y +$ genotype and we would expect a progeny ratio of 3 *yellow* : 1 wild type; the measured ratio is 33 : 12. This is certainly acceptable as a 3 : 1 ratio according to the χ^2 test and therefore consistent with our hypothesis. Is it consistent with the back-cross data?

From our hypothesis, wild-type individuals are $+ +$. *Yellow* phenotype may be either $Y +$ or YY. On this basis, some of the 33 F$_1$ *yellow* mice would be expected to be homozygous YY. If so, they could not give wild-type progeny, regardless of the genotype of the other parent. The data show all the F$_1$ *yellow* mice segregate wild-type progeny, and thus all must be heterozygous ($Y +$). Where, then, is the YY genotype?

Certain alleles of certain genes, when homozygous, are lethal in effect (so-called 'recessive lethal' alleles). Of course, dominant lethals would be eliminated from the population immediately, but recessive lethals can be maintained in a population in heterozygous individuals.

Let us return to the parental cross. If both parents are $Y +$, we would expect three progeny genotypes in a predictable ratio: namely $1YY : 2Y + : 1 + +$. If we now eliminate the YY potential progeny (lost *in utero*), we are left with a prediction of $2 Y +$ (*yellow*) : $1 + +$ (wild-type) surviving progeny. Is this consistent with the observation of 33 *yellow* and 12 wild-type progeny? A χ^2 test of these data against a 2 : 1 theoretical ratio gives a value of χ^2 of 0.9, which, with one degree of freedom, has an acceptable probability (>0.1). Thus our results are consistent with the 2 : 1 segregation ratio suggested by our hypothesis that *yellow* has a dominant effect on coat colour-masking the effect of the wild-type allele), but is a recessive lethal.

Is our revised hypothesis consistent with the back-cross results? We are now predicting that all *yellow* mice have a *Y +* genotype and that all wild-type mice are *+ +*. Any cross of *yellow* × *yellow* should give a 2 *yellow* : 1 wild-type progeny ratio, and the results of such back-crosses are consistent with this. Furthermore, any back-cross of *yellow* × wild type should give a ratio of 1 *yellow* : 1 wild type, again observed in the two corresponding back-crosses. Finally, any wild-type × wild-type cross should give only wild-type progeny, again observed.

The results therefore indicate that the original cross is between individuals heterozygous for a gene we have designated *Y* (for *yellow*), which is dominant over the wild-type allele in its effect on coat colour, but is lethal when homozygous.

Problem 6	Single gene segregation, sex-linkage

Question

In *Drosophila melanogaster*, there is a white-eyed mutant, which may be maintained as a true-breeding strain. Remember that wild-type flies have dark-red eyes. Two individual matings are made, in cross (i) between a white-eyed male and a wild-type red-eyed female, and in cross (ii) between a wild-type male and a white-eyed female, i.e. reciprocal crosses. The results are given below:

(i)   P   white male × red female
       F_1 red males, red females
       F_2 25% red males, 25% white males, 50% red females
(ii)   P   red male × white female
       F_1 white males, red females
       F_2 25% red males, 25% white females, 25% red females, 25% white females

What is the interpretation of these data?

Answer

As we have results in percentages, not real numbers, we cannot do statistical analysis on them. We must assume that they are taken from large enough samples to be meaningful.

From cross (i), ignoring the differential distribution of phenotypes in the two sexes in the F_2, we have a result that would be acceptable for a single gene segregating according to Mendelian principles, giving a 3 : 1 ratio of the dominant (red) to the recessive (white) phenotype. What is the cause of this non-random result with respect to progeny sex?

Drosophila melanogaster has four pairs of chromosomes in the female, three pairs being autosomes, the fourth pair being the X chromosomes (sex chromosomes). The male also has three pairs of autosomes but has only a single X chromosome paired with a Y chromosome, which is not present in normal females. If the white (*w*) gene was on any of the autosomes, it would have segregated in the F_2 independently of sex. The fact that *w* and sex did not show independence indicates that the *w* gene is sex-linked, i.e. it is on one of the sex chromosomes. Because it can be expressed in both males and females, it must be located on the X chromosome not the Y chromosome (although it is possible for a small number of male-specific genes to be located on the Y chromosome). For this particular eye colour gene, females should have two copies but males only one.

In cross (i), the F_1 flies are all red-eyed, from true-breeding red- and white-eyed parents respectively. The female F_1 flies should be heterozygous ($+w$) and are phenotypically red. This implies that the white eye allele (w) is recessive to its wild red counterpart.

We can now represent the cross diagrammatically:

Cross (i)	P		white male $X^w\,Y$		$\times$	red female $X^+\,X^+$	
	Gametes	X^w		Y	X^+		X^+
	F_1	$X^+\,X^w$		$X^+\,Y$	$X^+\,X^w$		$X^+\,Y$
		red		red	red		red
		female		male	female		male

F_2

		Female gametes	
		X^+	X^w
Male gametes	X^+	$X^+\,X^+$ red female	$X^+\,X^w$ red female
	Y	$X^+\,Y$ red male	X^w Y white male

In cross (ii), with the sexes and phenotypes reversed, the red male is $X^+\,Y$ and the white female from a true-breeding stock must be genotypically $X^w\,X^w$.

We can now represent the cross diagrammatically:

Cross (ii)	P		red male $X^+\,Y$		$\times$	white female $X^w\,X^w$	
	Gametes	X^+		Y	X^w		X^w
	F_1	$X^+\,X^w$		$X^w\,Y$	$X^+\,X^w$		$X^w\,Y$
		red		white	red		white
		female		male	female		male

F_2

		Female gametes	
		X^+	X^w
Male gametes	X^w	$X^+\,X^w$ red female	$X^w\,X^w$ white female
	Y	$X^+\,Y$ red male	$X^w\,Y$ white male

This appears to provide a satisfactory explanation for the data provided.

Problem 7	Cytoplasmic inheritance in mitochondria

Question

In *Neurospora crassa*, there is a slow-growing strain called *poky*. *Neurospora* has two mating types, *A* and *a*, and sexual reproduction depends on the crossing of an *A* strain with an *a* strain. If one parent is grown on crossing medium for several days until the female sexual initials (protoperithecia) develop, and it is then fertilized with conidia from the second parent, the first contributes the female gamete and the latter the male gamete. Although their nuclear genetic contributions are identical, the female parent contributes virtually all of the cytoplasm of the zygote.

The cross between wild type and *poky* can be made in reciprocal fashion, with either strain as the female parent. From such reciprocal crosses, the following results are obtained:

(i)	P	wild-type male × *poky* female
	F_1	all progeny are *poky*
(ii)	P	*poky* male × wild-type female
	F_1	all progeny are wild type

These results are obtained regardless of whether the wild-type strain is *A* and the *poky* strain is *a*, or vice versa. In all cases, there is a $1A : 1a$ segregation.

Explain these results.

Answer

The initial observation is that all progeny in both crosses have the same phenotype as the maternal parent. This is what is known as 'maternal inheritance'. As the chromosomal contribution of both parents is equal, it is unlikely that the relationships between nuclear genome contributions to the phenotype are going to be reversed depending on which is the female parent, and we must look elsewhere for an explanation. The fact that there is a $1 : 1$ segregation for the nuclear gene for mating type, with its two alleles *A* and *a*, confirms that behaviour of nuclear genes, and hence of the chromosomes that contain them, is normal.

An obvious difference between the two parental contributions is in cytoplasm, since, as stated in the question, the bulk of the cytoplasm of the zygote is derived from the female gamete. It is likely that the cytoplasm, not the nucleus, contains the hereditary determinant for wild-type or *poky* growth rate.

In all living organisms, DNA is the hereditary material. Are there any sites of DNA in the cytoplasm? DNA is found in the complex cytoplasmic organelles, mitochondria and chloroplasts. Since fungi are not photosynthetic plants, there are no chloroplasts. Therefore, mitochondria are the prime candidates as the cause for *poky* growth. In fact, the *poky* phenotype can be associated with a deficiency in the mitochondrial cytochromes, caused by a large deletion in the normal circular DNA molecule found in the mitochondria of wild-type *Neurospora*.

| Problem 8 | **Cytoplasmic inheritance in chloroplasts** |

Question

In the ornamental green plant *Mirabilis jalapa*, there exists a form with variegated leaves and stems, i.e. green areas (with cells containing chloroplasts) and yellow patches (cells without chloroplasts). Sectoring in a variegated *Mirabilis* plant gives rise to some shoots that are uniformly green and others that are uniformly yellow. The former grow well, and can be removed and propagated. The latter grow less well, being dependent on green parts of the plant for nutrition, and cannot be propagated. Green shoots stay uniformly green and yellow shoots stay uniformly yellow, but variegated shoots may subsequently sector further green or yellow shoots.

Flowers may develop on green, yellow and variegated shoots. If all possible intercrossings are made (pollen from flowers on each of the three types of shoot used to fertilize ova in flowers on each of the three) the following results are obtained:

		Female gamete on shoot type		
		green	yellow	variegated
	green	green	yellow	green yellow variegated
Pollen from	yellow	green	yellow	green yellow variegated
	variegated	green	yellow	green yellow variegated

Explain these results.

Answer

The stability of the green and yellow phenotypes implies that they have become different in a stable and heritable way, by the somatic segregation of some hereditary determinant. We will therefore treat them as genetically distinct types.

In the cases of either green or yellow female parental shoots, all the progeny resemble the female parental phenotype. This is clearly a form of maternal inheritance and suggests that the genetic determinants are located in the cytoplasm. As it clearly affects the presence or absence of mature geeen chloroplasts, we may postulate that some difference in the chloroplast DNA is responsible.

All plant cells contain chloroplast initials, which all contain the standard circular chloroplast DNA. Only in the normally green tissues of the plant do these initials develop the mature structure and chlorophyll content of mature chloroplasts. It is in the form of chloroplast initials, not mature chloroplasts, that the ability to make chloroplasts is transmitted through the cytoplasm of the egg into the plants of the next generation.

Eggs of flowers on green shoots contain only normal chloroplast initials and therefore their seeds produce only green offspring. Eggs of flowers on yellow shoots contain only mutant chloroplast initials and therefore their seeds produce only yellow seedlings (which die as soon as the food reserves of the seed are exhausted, as they cannot photosynthesize more). Flowers on variegated shoots are capable of segregating either single chloroplast initial type into eggs, as well as maintaining some with both types of initial. Hence, flowers on variegated shoots produce three types of progeny, green, yellow and variegated. These results indicate also that there is typically more than one chloroplast initial per egg.

Problem 9

Assortment of two unlinked genes in ordered tetrads

Question

In the ascomycete fungus *Sordaria fimicola*, several genes affect ascospore colour. Wild-type ascospores (+) are dark green in colour, and single mutant strains may have buff (*b*), yellow (*y*), grey (*g*) or white (*w*) ascospores. A *b* mutant strain is crossed with a *y* mutant strain and the following ordered tetrads (NB each member of the tetrad is represented by a pair of ascospores in the ascus) were obtained from the cross:

	i	ii	iii	iv	v	· vi	vii	viii	ix	x
	18	22	10	9	8	13	5	6	4	5
pair 1	*y*	+	*w*	*w*	*y*	*y*	*w*	*w*	*b*	*b*
pair 2	*y*	+	*y*	*y*	*w*	*w*	*b*	*b*	*w*	*w*
pair 3	*b*	*w*	*b*	+	+	*b*	*y*	+	+	*y*
pair 4	*b*	*w*	+	*b*	*b*	+	+	*y*	*y*	+

Are the *w* and *b* mutants allelic? What is the origin of the white ascospores in this cross? Map the gene(s) and centromere(s) segregating in the cross.

Answer

If *y* and *b* were allelic (i.e. mutant in the same gene), there would be little if any recombination between them, so almost all asci would contain 4 *y* and 4*b* spores. Such an ascus segregation is termed a *parental ditype* (PD), as it contains only two phenotypes, with four spores of each type, and each of the two types resembling one of the two parents, either buff or yellow. However, only 18 out of the 100 asci are of this type. The two parents must therefore be mutant in different genes.

Apart from the PD ascus segregation type, there are two others. One of these again has four spores of each of two phenotypes (hence a ditype), but neither resembles either parent. In this case, the phenotypes are wild (dark) and white. Because this is a ditype ascus but neither type is parental, it is called a *non-parental ditype* (NPD). The third type of ascus has two spores of each of four phenotypes and is therefore called a *tetratype* or TT. In a TT, two of the spore pairs resemble the two parental phenotypes (buff and yellow in this case) and two are non-parental (dark and white).

If the wild-type alleles of the *b* and *y* genes produce different pigments, and the wild-type spore colour is dark because both are present, we can postulate that a double mutant (b y) will have neither pigment and therefore be colourless (white). Hence, although the question states that single mutant w strains are known, that is not what we have in this case.

With the two genes, b and y, segregating in this cross, a PD ascus would contain 4 b spores and 4 y spores, an NPD ascus 4 wild-type ($+$) and four white (b y) spores and a TT ascus 2 $+$, 2 b, 2 y and 2 white (b y) spores.

The two genes, b and y, might be either linked or unlinked. What evidence do we have relating to this? For unlinked genes, the PD has the same frequency as the NPD. In this case, there are 18 PD (class i) and 22 NPD (class ii), statistically equal by a simple χ^2 test. For genes to be demonstrably linked, the PD must be greater than the NPD. We will continue with the hypothesis that the two genes are unlinked and hence independently assorting.

Can we now calculate the distances of the b and y genes from their respective centromeres? From the hypothesis that white spores are b y double mutants, we can. Taking first b, it is segregating at the first division in classes i, ii, vii, viii, ix and x ($17 + 23 + 5 + 6 + 4 + 5$) and at the second division in classes iii, iv, v and vi ($10 + 9 + 8 + 13$). Using the relationship that the distance from the centromere is one half of the frequency of second-division segregation tetrads (expressed as a percentage), centromere distance of b is:

$$\frac{1}{2} \times \frac{40}{100} \times 100\%$$

$$= 20 \text{ map units}$$

Analysing the segregation of y in the same way, first-division segregations are seen in classes i, ii, iii, iv, v and vi ($17 + 23 + 10 + 9 + 8 + 13$) and second-division segregations are seen in classes vii, viii, ix and x ($5 + 6 + 4 + 5$). Therefore centromere distance of y is:

$$\frac{1}{2} \times \frac{20}{100} \times 100\%$$

$$= 10 \text{ map units}$$

Thus we have determined that the two genes are unlinked, that the double mutant phenotype is white and that b is 20 map units from its centromere and y 10 map units from its centromere. If we had only had random meiotic progeny from the same cross, we could have only deduced the double mutant phenotype and the fact that b and y were unlinked.

| Problem 10 | **Two unlinked genes in haploid, random progeny** |

| Question | In *Saccharomyces cerevisiae*, a cross is made between two non-allelic mutant strains, *ura3* and *ura4*, both of which require uracil for growth. In 100 meiotic progeny, 28 are wild type and 72 require uracil for growth. Explain these results. |

Answer

We are given that *ura3* and *ura4* are not allelic, i.e. they are mutant at two different genes. We may therefore conclude that the *ura3* strain has a wild-type allele (*+*) at the *URA4* locus and the *ura4* strain has a wild-type allele (*+*) at the *URA3* locus.

The cross is therefore:

$$ura3 \; + \; \times \; + \; ura4$$

Be careful always to list the genes in the same order; in this case the allele of the *ura3* gene, as above, is always to the left of *ura4*, i.e. the first strain above is mutant at *ura3* but has the wild-type allele for *ura4*, whereas the second is wild type for *ura3* but mutant at *ura4*.

From this cross, we would expect four progeny genotypes:

ura3 +	*ura3 ura4*	*+ura4*	*+ +*
all three types auxotrophic			prototrophic

As you will observe, because the two mutants have the same phenotype, as does the double mutant, we may presume only one of the recombinant genotypes is phenotypically distinct. However, although we are losing some information that we would have had if the phenotypes had differed, we can still interpret that information we do have.

If the two genes were unlinked, all four genotypes would occur at equal frequency, so 25% would be in the prototrophic phenotype class. If linked, significantly less than 25% would be prototrophs.

To test for independent assortment, we must carry out a χ^2 test of the actual results (72 and 28) against an ideal 3:1 ratio (75:25 for the total of 100 progeny):

$$\chi^2 = \frac{(70-75)^2}{75} + \frac{(28-25)^2}{25}$$

$$= \frac{9}{75} + \frac{9}{25} = \frac{36}{75} = 0.48$$

This χ^2 value of 0.48, for one degree of freedom, gives a probability value of about 0.5, indicating that the results are compatible with 1 in 4 progeny being prototrophic (and thus of the *+ +* recombinant genotype). This is consistent with the hypothesis of independent assortment.

A supplementary question could ask how it might be possible to distinguish between the three auxotrophic genotypes, *ura3 +*, *ura3 ura4* and *+ ura4*. This could be achieved in various ways, but the simplest will be described.

Any auxotrophic progeny strain should be test-crossed to each of the two singly mutant types, *ura3 +* and *+ ura4*. A *ura3 ura4* double mutant will fail to give any wild-type progeny from either cross. A *ura3 +* progeny strain will only give wild-type progeny in a cross with a *+ura4* tester and a *+ura4* progeny strain only with a *ura3 +* tester, as shown below:

Unknown	Tester	Progeny genotypes	
ura3 +	*ura3 +*	all *ura3 +*	all auxotrophic
ura3 +	*+ura4*	$\frac{1}{4}$ *ura3 +*, $\frac{1}{4}$ *+ura4*, $\frac{1}{4}$ *ura3 ura4*, $\frac{1}{4}$ *++*	$\frac{1}{4}$ prototrophic
+ura4	*ura3 +*	$\frac{1}{4}$ *ura3 +*, $\frac{1}{4}$ *+ura4*, $\frac{1}{4}$ *ura3 ura4*, $\frac{1}{4}$ *++*	$\frac{1}{4}$ prototrophic
+ura4	*+ura4*	all *+ura4*	all auxotrophic
ura3 ura4	*ura3 +*	$\frac{1}{2}$ *ura3 ura4*, $\frac{1}{2}$ *ura3 +*	all auxotrophic
ura3 ura4	*+ura4*	$\frac{1}{2}$ *ura3 ura4*, $\frac{1}{2}$ *+ura4*	all auxotrophic

Problem 11 — Two unlinked genes in diploid with dominance

Question

In this question, we will look at the kind of data Gregor Mendel would have obtained in his monastery garden. It involves, of course, the garden pea, *Pisum sativum*, and two rather different varieties, one producing seeds with rounded yellow cotyledons and the other seeds with wrinkled green cotyledons.

In a cross between these two true-breeding varieties, all the progeny seeds (the start of the F_1 generation) were round and yellow. These seeds were grown into adult F_1 plants and allowed to self-pollinate to produce seeds (the start of the F_2 generation). In a large sample of this F_2 population, both seed shape and seed colour was segregating. In 320 F_2 seeds classified, there were:

171 round yellow
 65 round green
 60 wrinkled yellow
 24 wrinkled green

Interpret these results. Also, predict the results of crossing an F_1 plant to a plant of the parental, wrinkled green variety.

Answer

It is possible to make assumptions and achieve a few short cuts in this answer, but let us take a methodical approach. Firstly, we have two pairs of alternative characters, one affecting cotyledon shape and the other affecting cotyledon colour. Let us consider these separately at first.

For shape, crossing round with wrinkled (the P generation) gave an F_1 that were all round. In the F_2, there were 236 round and 84 wrinkled, i.e. a 3 : 1 ratio in the F_2, verifiable by the χ^2 test. This pattern of behaviour is consistent with the segregation of two alleles of a gene affecting cotyledon shape, with round dominant over wrinkled.

For convenience, let us designate the allele determining roundness W and the wrinkled allele w. The initial cross is round ($W\,W$) × wrinkled ($w\,w$), the F_1 are all round ($W\,w$) and in the F_2 there is a 1 $W\,W$: 2 $W\,w$: 1 $w\,w$ genotypic segregation. The F_2 phenotypic segregation is of course 3 round (1 $W\,W$ and 2 $W\,w$) to 1 wrinkled ($w\,w$).

The same applies to cotyledon colour, where a yellow × green cross produces yellow F_1, and segregates 231 yellow and 89 green in the F_2 (again an acceptable 3 : 1 ratio). Yellow is dominant over green, and we can designate the alleles G for yellow and g for green.

If the two genes are assorting independently and the two alleles of each gene are randomly associating in the production of gametes from the F_1 plants, we can predict the ratio in the F_2. Let us do this as part of a diagram of the whole pedigree:

P	phenotypes	round yellow	×		wrinkled green
	genotypes	$W\,W\,G\,G$			$w\,w\,g\,g$
	gametes	$W\,G$			$w\,g$

F_1	phenotype		round yellow		
	genotype		$W\,w\,G\,g$		

	gamete ratio	1	1	1	1
	genotypes	$W\,G$	$W\,g$	$w\,G$	$w\,g$

The consequences of random combinations of these four equal types in fertilization to create the F_2 is conveniently shown in the chequerboard diagram below (the Punnett square):

		Male gametes			
		$W\,G$	$W\,g$	$w\,G$	$w\,g$
	$W\,G$	$W\,W\,G\,G$ round yellow	$W\,W\,G\,g$ round yellow	$W\,w\,G\,G$ round yellow	$W\,w\,G\,g$ round yellow
	$W\,g$	$W\,W\,G\,g$ round yellow	$W\,W\,g\,g$ round green	$W\,w\,G\,g$ round yellow	$W\,w\,g\,g$ round green
Female gametes	$w\,G$	$W\,w\,G\,G$ round yellow	$W\,w\,G\,g$ round yellow	$w\,w\,G\,G$ wrinkled yellow	$w\,w\,G\,g$ wrinkled yellow
	$w\,g$	$W\,w\,G\,g$ round yellow	$W\,w\,g\,g$ round green	$w\,w\,G\,g$ wrinkled yellow	$w\,w\,g\,g$ wrinkled green

Of the 16 equal possible combinations in the above grid:

9 are round yellow (1 $WWGG$, 2 $WWGg$, 2 $WwGG$, 4 $WwGg$)
3 are round green (1 $WWgg$, 2 $Wwgg$)
3 are wrinkled yellow (1 $wwGG$, 2 $wwGg$)
1 is wrinkled green ($wwgg$)

The actual data give 171, 65, 60 and 24 progeny in these four classes respectively, which can be tested against the expected results for independent assortment in a sample of that size (180, 60, 60 and 20) by the χ^2 test, this time with three degrees of freedom as we have four classes:

$$\chi^2 = \frac{(171-180)^2}{180} + \frac{(65-60)^2}{60} + \frac{(60-60)^2}{60} + \frac{(24-20)^2}{20}$$

$$= \frac{81}{180} + \frac{25}{60} + \frac{0}{60} + \frac{16}{20}$$

$$= 1.666$$

A χ^2 value of 2.12 with three degrees of freedom gives $P > 0.5$, so the results are consistent with independent assortment.

The second part of the question asks you to predict the result of a cross of an F_1 plant (of $WwGg$ genotype) to its wrinkled green parent ($wwgg$). This is a back-cross to a strain homozygous recessive for both genes under study. The F_1 plant (of genotype $WwGg$) will produce, by independent assortment as determined above, the four gamete types WG, Wg, wG and wg in equal frequencies. Each type will combine with a wg gamete from the back-cross parent, the two recessive alleles having no effect on the phenotype of the progeny. Therefore, the back-cross progeny phenotypes will be a direct reflection of the gamete genotypes produced by the F_1 plant. We therefore expect the ratio, genotypes and phenotypes of the progeny of this back-cross to be:

1	1	1	1
$WwGg$	$Wwgg$	$wwGg$	$wwgg$
round yellow	round green	wrinkled yellow	wrinkled green

Problem 12	**Two unlinked genes in diploid with interaction**

Question

In *Drosophila melanogaster*, wild-type flies have dark-red eyes. A fly from a true-breeding strain with scarlet (bright-red) eyes was crossed to a fly from another true-breeding strain with brown eyes. The F_1 flies were all wild type for eye colour. When these F_1 flies were inter-crossed, the F_2 flies were classified as follows:

472 wild-type (dark-red eyes)
138 scarlet eyes
145 brown eyes
 45 white eyes

How many genes are involved? If more than one gene is involved, are any of them linked? What genotype(s) are the F_2 white-eyed flies? Draw out the full cross, indicating genotypes in each generation and genotypes of the gametes produced by the P and F_1 generations.

Answer

A good starting point here, but not the only possible one, is the F_2. There are four phenotypic classes with numbers of 472, 138, 145 and 45. That should suggest the possibility of a $9:3:3:1$ ratio. To test this hypothesis, a χ^2 test should be done against the expectation, in a sample of 800, of 450, 150, 150 and 50. The χ^2 value is approximately 2.65, which, for the three degrees of freedom in this case, gives $P > 0.5$, clearly an acceptable probability.

A $9:3:3:1$ ratio is indicative of two independently assorting genes segregating from a doubly heterozygous F_1. Hence we have two genes involved, which are unlinked. There is no mention of any sex difference, so we can ignore the possibility of sex linkage.

Returning to the P and F_1 generations, a scarlet parent and a brown parent give all wild-type F_1 flies. Both brown (*b*) and scarlet (*s*) are therefore recessive to their wild-type alleles *B* and *S*.

The $9:3:3:1$ F_2 ratio implies that all the F_1 flies were heterozygous for both genes (*B b S s*). For this to be the case, the two P generation flies must have been homozygous for both genes, so only one gamete type would come from each parent.

The scarlet parent must have been *s s* in genotype. The brown parent must have been *b b* in genotype. What were they each with respect to the other gene? The scarlet parent must have been the source of the *B* allele in the F_1 and was therefore *B B*. Likewise, the brown parent must have been the source of the *S* allele in the F_1 and was therefore *S S* in genotype. To give their genotypes in full, the scarlet parent was *B B s s* and the brown parent was *b b S S*. Thus the former contributed a *B s* gamete to the F_1 and the latter a *b S* gamete.

In the F_2, using this hypothesis, we would predict one-sixteenth of the flies to have a *b b s s* genotype. In the data, one-sixteenth have a white-eye phenotype. This is consistent with the white-eyed F_2 flies lacking the scarlet pigment by being *s s* and the brown pigment by being *b b*, therefore giving white eyes through lack of any eye pigment. Conversely, the dark-red eyes of wild-type flies results from having both the scarlet and the brown pigment present. Therefore, all white-eyed F_2 flies have the same genotype, *b b s s*.

The diagram of the whole experiment is therefore:

P	phenotypes	scarlet-eyed		×		brown-eyed
	genotypes	$BBss$				$bbSS$
	gametes	Bs				bS
F_1	phenotype			wild type		
	genotype			$BbSs$		
	gamete ratio	1	1		1	1
	genotypes	BS	Bs		bS	bs

The consequences of random combinations of these four equal types in fertilization to create the F_2 is conveniently shown in the Punnett square below:

Sperm genotypes from F_1

		BS	Bs	bS	bs
	BS	$BBSS$ wild type	$BBSs$ wild type	$BbSS$ wild type	$BbSs$ wild type
	Bs	$BBSs$ wild type	$BBss$ scarlet	$BbSs$ wild type	$Bbss$ scarlet
Egg genotypes from F_1	bS	$BbSS$ wild type	$BbSs$ wild type	$bbSS$ brown	$bbSs$ brown
	bs	$BbSs$ wild type	$Bbss$ scarlet	$bbSs$ brown	$bbss$ white

Problem 13	**Two unlinked genes in diploid, one autosomal, one sex-linked**

Question

In *Drosophila melanogaster*, a white-eyed male is crossed to a wild-type virgin female. All of their progeny are wild type. The F_1 flies are inter-crossed, and the F_2 comprises the following phenotypes and frequencies:

80 wild-type females
25 brown-eyed females
34 wild-type males
14 brown-eyed males
47 white-eyed males

Explain these results fully.

Answer

In a very short question, we clearly have some rather complex data. Let us start with considering whether a fly has white or coloured eyes and delay consideration of the actual colour until later.

In 200 flies, we have 153 with coloured eyes and 47 with white eyes (lacking any coloured pigments). This can be tested with the χ^2 test and shown to be an acceptable $3:1$ ratio, consistent with segregation of two alleles, W producing eye pigment and w

not doing so. A complication is that all the white-eyed flies are male (105 coloured females, 48 coloured males and 47 colourless males). This is an indication that the gene is sex-linked. The original white parent is of genotype X^w Y and the female parent $X^W X^W$. The F_1 males were therefore X^W Y and the females $X^W X^w$.

Now we have the complicating problem of the brown-eyed flies in the F_2. As the white-eye phenotype masks the expression of any eye pigment, we cannot classify the F_2 white-eyed males as either brown or red. However, in the F_2 females, we have an acceptable $3:1$ ratio of wild type (red-eyed) to brown-eyed (80:25). Within the non-white F_2 males, again there is an acceptable $3:1$ segregation of wild type to brown-eyed (34:14) and no suggestion of sex-linkage of the brown-eye character. Test both of these segregations with the χ^2 test.

The simplest hypothesis is that we have two genes segregating in this experiment. They are wild-type (W) and white (w) alleles of a sex-linked gene, and wild-type (B) and brown (b) alleles of an autosomal gene. As all the F_1 flies are wild type, the wild-type female parent must have been homozygous wild type at both genes ($X^W X^W B B$). The white-eyed male parent must have had the white allele (w) on its single X chromosome and also have been homozygous recessive for the brown-eye gene ($b b$). The $b b$ phenotype was masked because it was also phenotypically white. This relationship, where one mutant phenotype masks that normally associated with a second, has a special name, *epistasy*. The white-eye phenotype is said to be epistatic to the brown-eye phenotype.

From this hypothesis, can we predict the F_2 segregation and is it compatible with the observed results?

	P	phenotypes	white male	×	wild-type female
		genotypes	$X^w Y\, b b$		$X^W X^W B B$
		gametes	$X^w b$ $Y b$		$X^W B$
	F_1	phenotype	wild-type males		wild-type females
		genotypes	$X^W Y B b$		$X^W X^w B b$

Gametes from F_1 females

		$X^W B$	$X^w B$	$X^W b$	$X^w b$
	$X^W B$	$X^W X^W B B$ wild-type female	$X^W X^w B B$ wild-type female	$X^W X^W B b$ wild-type female	$X^W X^w B b$ wild-type female
	$X^W b$	$X^W X^W B b$ wild-type female	$X^W X^w B b$ wild-type female	$X^W X^W b b$ brown-eye female	$X^W X^w b b$ brown-eye female
Gametes from F_1 males	$Y B$	$X^W Y B B$ wild-type male	$X^w Y B B$ white-eye male	$X^W Y B b$ wild-type male	$X^w Y B b$ white-eye male
	$Y b$	$X^W Y B b$ wild-type male	$X^w Y B b$ white-eye male	$X^W Y b b$ brown-eye male	$X^w Y b b$ white-eye male

Therefore the prediction from our hypothesis is a ratio of: 6 red female : 2 brown female : 3 red male : 1 brown male : 4 white male. For our 200 scored F_2 progeny, we would therefore expect in these five classes 75, 25, 37.5, 12.5 and 50 respectively. With these expected numbers and the actual data, we can carry out a χ^2 test, with four degrees of freedom as we have five phenotypic classes of F_2 progeny. With a calculated value of χ^2 of approximately 0.9, we have a probability of about 0.9. This indicates that our hypothesis is acceptable.

Problem 14

Two linked genes, unordered tetrad analysis

Question

In the yeast *Saccharomyces cerevisiae*, a cross is made between a wild-type strain and a double mutant strain *ura* (requires a pyrimidine, e.g. uracil, for growth) *met* (requires methionine for growth). In 200 unordered tetrads isolated, tested and classified, the following three types were obtained:

i		ii		iii	
100 tetrads		92 tetrads		8 tetrads	
ura	+	*ura*	+	*ura*	*met*
ura	+	*ura*	*met*	*ura*	*met*
+	*met*	+	+	+	+
+	*met*	+	*met*	+	+

Are the two genes linked or unlinked? If they are linked, what is the distance between them?

For a genetic interval, it is possible to have zero, one, two or even more crossovers occurring. Draw the arrangement of crossovers in non-crossover, single crossover and all possible double crossover tetrads as they would be in the pachytene stage of meiotic prophase I.

Answer

The determination of linkage or independent assortment depends on the relative frequencies of PD and NPD. If detectably linked, PD must be greater than NPD. Note that it is possible for two 'linked' genes to be so far apart that the percentage recombination is so high (about 50%) that it is not readily distinguished from independent assortment. In this case, PD is class i (100 tetrads) and NPD is class iii (8 tetrads), so there is no difficulty in deciding that they are linked. If the NPD value were closer to the PD, a χ^2 test should be applied to determine statistically whether the two values are significantly different.

Having determined that the two genes are linked, can we calculate the distance they are apart?

For two linked genes, the PD ascus class represents no detectable crossovers between the two genes in the ascus, as the two progeny types contain the parental combinations of alleles of the two genes segregating and thus require no crossing over. The TT class requires only a single detectable crossover between the two segregating genes between two of the four chromatids, resulting in the necessary two parental and two recombinant progeny types within the ascus. In NPD segregation, all four chromatids are recombinant between the two marker genes, needing a four-strand double crossover between them to generate all recombinant progeny. Of course, two crossovers in the defined interval involving the same chromatid(s) can cancel out, leaving some crossovers undetected. This only becomes a major factor in long genetic intervals, where multiple crossovers become more common.

Hence half the chromatids in a TT ascus are necessarily recombined in the interval between the genes, and all of the chromatids in the NPD asci.

Therefore, the distance between the genes is:

$$\frac{NPD + \frac{1}{2}TT}{PD + TT + NPD} \times 100\%$$

Returning to the data, class ii (TT) asci have a single crossover involving two of the four chromatids, and class iii (NPD) asci have two crossovers. Substituting in the equation, the distance between *ura* and *met* is:

$$\frac{8 + (\frac{1}{2} \times 92)}{100 + 92 + 8} \times 100\%$$

$$= \frac{8 + 46}{200} \times 100\% = \frac{54}{2}$$

$$= 27 \text{ map units}$$

The possible types of non-crossovers (NCO), single crossovers (SCO) and double crossovers (DCO) in an interval between two linked genes are shown below:

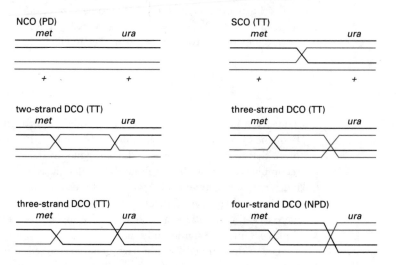

The genetic consequences of multiple crossovers in an interval are very interesting. A two-strand DCO has one crossover cancelling out the effect of the other, resulting in no detectable crossing over in the interval. Three-strand DCO are indistinguishable from single crossovers. As multiple crossovers reduce estimates of genetic distance, the error increasing with distance because the probability of multiple crossing over increases with distance, estimates of longer genetic intervals become increasingly unreliable. The use of correction factors (mapping functions) attempts to minimize this problem, but with questionable results.

Problem 15

Three genes, two linked, one unlinked, centromere markers

Question

Here is another problem on tetrad analysis, this time involving three genes in *Saccharomyces cerevisiae*: *ade* (requirement for adenine), *his* (histidine) and *tyr* (tyrosine). The actual cross is *ade + tyr* × *+ his +*. Being yeast, the tetrads are unordered. The results obtained were:

class i			class ii			class iii		
2			18			28		
ade	+	+	ade	+	tyr	ade	+	tyr
ade	+	+	ade	+	+	ade	+	tyr
+	his	tyr	+	his	tyr	+	his	+
+	his	tyr	+	his	+	+	his	+

class iv			class v			class vi		
4			16			32		
ade	his	+	ade	his	tyr	ade	his	tyr
ade	his	+	ade	his	+	ade	his	tyr
+	+	tyr	+	+	+	+	+	+
+	+	tyr	+	+	tyr	+	+	+

Which genes if any are linked? What is or are the distances(s) between these genes? Although unordered tetrads, can anything be deduced about the distances of any of the genes from their centromere(s)?

Answer

When confronted with data involving the segregation of multiple genes, always consider them a pair at a time.

For *ade* and *his*, the cross was *ade +* × *+ his*. Therefore classes iv, v and vi are PD (4 + 16 + 32 = 52) and classes i, ii and iii are NPD (2 + 18 + 28 = 48). PD equal NPD so *ade* and *his* are unlinked. There are no TT asci. As a TT ascus requires one gene to segregate at the first division and the other at the second division, it can be deduced that in no ascus is either gene segregating at the second division. Therefore the two genes are so closely linked to their centromeres that neither has any recombination in that interval in the sample of 100 asci. Therefore, each has a centromere distance of 0 map units.

For *ade* and *tyr*, the cross was *ade tyr* × *+ +*. Therefore classes iii and vi are PD (28 + 32), classes i and iv are NPD (2 + 4) and ii and v are TT (18 + 16). PD and NPD are statistically not equal, so we have evidence of linkage between the two genes. Their distance apart is:

$$\frac{\text{NPD} + \frac{1}{2}\text{TT}}{\text{PD} + \text{TT} + \text{NPD}} \times 100\%$$

Substituting in this case, the distance from *ade* to *tyr* is:

$$\frac{6 + \left(\frac{1}{2} \times 34\right)}{60 + 34 + 6} \times 100\%$$

$$= \frac{6 + 17}{100} \times 100$$

$$= 23 \text{ map units}$$

As we know that *ade* is so closely linked to its centromere that it has not recombined in any of the 100 asci, it may be regarded as a 'centromere marker'. Therefore, *tyr* is 23 map units from not only *ade* but also its centromere.

Now let us look at *his* and *tyr*, the cross being *his* + × + *tyr*. For these two genes, PD are iii + iv (28 + 3), NPD are i + vi (1 + 32) and TT are ii + v (18 + 16). We can show that *his* and *tyr* are unlinked, as PD (31) and NPD (33) are statistically equal by the χ^2 test. However, we already know that *ade* and *his* are unlinked and that *ade* and *tyr* are linked at a distance of 23 map units, so this merely confirms what we should know already. If we look at the classes in the TT category, these are the same two classes as for *ade* and *tyr*. This should not come as a surprise, as we know now that both *ade* and *his* are only giving first-division segregations, so any TT are due entirely to the second-division segregations of *tyr*.

This is a very useful observation. Once any gene is mapped and shown to be very closely linked to a centromere, the centromere distance of any other gene, on any chromosome, can be determined by crossing it to the 'centromere' marker, determining the TT frequency, i.e. the second-division segregation frequency of the unmapped gene, and doing the relevant calculation (half of the second-division segregation frequency).

It is possible to calculate centromere distances from unordered tetrad data without reference to a 'centromere marker', but this involves a rather more complicated calculation. However, it is worth noting the possibility of doing this for either three unlinked genes or two linked genes on opposite chromosome arms and a third unlinked gene.

Problem 16	Three genes, two linked, one unlinked in haploid

Question

In *Aspergillus nidulans*, a strain requiring proline and alanine for growth was crossed to a strain requiring tyrosine for growth. The progeny from this cross were:

 7 required proline, alanine and tyrosine
 5 required proline and tyrosine
35 required proline and alanine
45 required proline
 8 required alanine
48 required alanine and tyrosine

42 required tyrosine

10 required no supplement

Interpret these results.

Answer

Taking the amino acid requirements one at a time, there are 92 that require proline and 108 that do not. This is, by the χ^2 test, an acceptable $1:1$ ratio, consistent with the cross involving two alleles of a gene involved in proline metabolism (*pro* and *+*). The results are similar for alanine (98 requiring alanine and 102 not), so again two alleles are involved (*ala* and *+*); similarly for tyrosine (102 tyrosine requiring (*tyr*) and 98 not (*+*)).

Let us now take the genes two at a time and look for independent assortment or linkage. Considering first *pro* and *ala*, the four possible genotypes and their frequencies are:

pro	*ala*	$(7 + 35 = 42)$
pro	*+*	$(5 + 45 = 50)$
+	*ala*	$(8 + 48 = 56)$
+	*+*	$(10 + 42 = 52)$

This is acceptable by the usual test as a $1:1:1:1$ ratio, so the *pro* and *ala* genes appear to be unlinked.

Now isolating *pro* and *tyr*, the genotypes and numbers are:

pro	*tyr*	$(7 + 5 = 12)$
pro	*+*	$(35 + 45 = 80)$
+	*tyr*	$(48 + 42 = 90)$
+	*+*	$(8 + 10 = 18)$

This is clearly incompatible with independent assortment, being very far removed from four equal frequencies. The two parental genotypes are clearly in excess, indicating linkage between these two genes. Calculating the percentage recombination, and hence the map distance is simple, using the standard equation:

$$\frac{\text{no. of recombinants}}{\text{total progeny}} \times 100\%$$

$$= \frac{12 + 18}{200} \times 100$$

$$= 15 \text{ map units}$$

We would now expect that *ala* and *tyr* are unlinked, but we should check to be sure:

ala	*tyr*	$(7 + 48 = 55)$
ala	*+*	$(8 + 35 = 43)$
+	*tyr*	$(5 + 42 = 47)$
+	*+*	$(10 + 45 = 55)$

As expected, this is acceptable as a $1:1:1:1$ ratio, consistent with independent assortment.

To summarize, the genotypes of the two strains are:

pro   +   ala   +   tyr   +

and

The genetic map, using the gene symbols to represent the genes rather than their specific alleles, is:

pro   tyr   ala

└──── 15 ────┘

Problem 17

Classic three-point cross in haploid, random progeny

Question

In *Neurospora crassa*, the double mutant strain *pyr-1* (pyrimidine requiring) *met-5* (methionine requiring) is crossed to the strain *ade-6* (adenine requiring). Progeny are isolated, germinated and grown on complete medium, and subsequently tested for the three possible biochemical requirements. The progeny types and numbers are:

pyr	ade	met	3
pyr	ade	+	71
pyr	+	met	358
pyr	+	+	40
+	ade	met	49
+	ade	+	382
+	+	met	77
+	+	+	0

Determine the linkage or independent assortment of the genes involved. If any genes are linked, what are the distances between them? Draw a genetic map showing the three genes. If possible, determine the type and extent of any crossover interference.

Answer

In any question involving more than two genes, it is necessary to list the genes in some kind of order. One cannot assume any relationship of this arbitrary order with any linkage order of the genes. This genetic order must be determined from the numerical data. Likewise, there is no reason to assume any particular combination of alleles such as one parent having all wild-type alleles and the other all mutant alleles. Again the particular combinations of alleles must be determined from the actual data.

There are several ways of tackling this problem. It is possible to use an approach employed in several earlier problems. Consider the genes a pair at a time (*pyr* with *ade*, *pyr* with *met*, *ade* with *met*) and look for inequalities in frequencies of the four possible genotypes as evidence of linkage:

pyr	+	398	pyr	+	111	ade	+	451
pyr	ade	74	pyr	met	361	ade	met	52
+	+	77	+	+	382	+	+	40
+	ade	431	+	met	126	+	met	435

This indicates that all three genes are linked. Using the standard method for calculating percentage recombination, the distances between the pairs of genes are:

$$pyr - ade = \frac{74 + 77}{1000} \times 100\%$$

$$= \frac{151}{10} = 15.1 \text{ map units}$$

$$pyr - met = \frac{111 + 126}{1000} \times 100\%$$

$$= \frac{237}{10} = 23.7 \text{ map units}$$

$$ade - met = \frac{52 + 40}{1000} \times 100\%$$

$$= \frac{92}{10} = 9.2 \text{ map units}$$

From the three distances, we can conclude that *ade* is between *pyr* and *met* on the necessarily linear order of the genes. On this basis, we can construct a map of the linkage group (chromosome) on which the three genes are located, drawn below to show the gene order and inter-gene distances:

Of course, the mirror image of this would also be correct, as there is nothing to say which gene, *pyr* or *met*, is at which end.

The final part of the question asks if there is any evidence of crossover interference. Interference is detected when the number of DCO observed (in this case the genotypes + + + and *pyr ade met*) differ from what would be predicted as the product of the individual crossover frequencies in the two adjacent intervals defined by the three linked genes. There is 15.1% recombination in the *pyr–ade* interval and 9.2% recombination in the *ade–met* interval, so the expectation in a sample of 1000 progeny is 13.892 (15.1% × 9.2% = 1.3892%) DCO. Only three are observed.

If the observed DCO frequency is less than expected, interference is positive; if more than expected, interference is negative. Interference is generally positive, as it is in this case. The extent of that interference is estimated by dividing the observed number of DCO by the expected number, to give the *coefficient of coincidence*. In this example:

$$\text{coefficient of coincidence (C)} = \frac{3}{13.892} \simeq 0.215$$

$$\text{Interference} = 1 - C = 0.785$$

That fully answers the question.

It is actually possible to determine that the three genes are linked, and also to determine the gene order, without doing a single calculation. This is based on the fact that if the frequency of recombinant genotypes is less than 50% the genes involved are linked, and if they equal 50% there is no basis for claiming linkage (of course, there is no theoretical basis to expect statistically more than 50% recombinants).

Consider three linked genes, A, B and C, defining two intervals, I and II. The recombination frequency in interval I is x, so the frequency of non-recombinants is $(1-x)$. In interval II, recombination frequency is y, so non-recombination frequency is $(1 - y)$. Being linked genes, $x < 0.5$ and $y < 0.5$.

Type			Frequency
NCO	$(1 - x)$	$(1 - y)$	$(1 - x)\,(1 - y)$
SCO in I	x	$(1 - y)$	$x(1 - y)$
SCO in II	$(1 - x)$	y	$y(1 - x)$
DCO (I, II)	x	y	xy

Each category above (NCO, SCO I, SCO II and DCO) will of course be represented in the results by two reciprocal genotype classes.

It is clear that sum of the two NCO classes, being the product of two frequencies each > 0.5, will be the most frequent, and the sum of the two DCO classes, being the product of two frequencies each < 0.5, will be the two smallest classes. The four SCO classes will have intermediate values.

Looking back to the data in the question, the two largest (hence NCO) classes are *pyr + met* and *+ ade +*, and the two smallest are *pyr ade met* and *+ + +*.

Let us consider a hypothetical case of three linked genes, *r*, *s* and *t*, segregating in a cross. The gene order is *r–s–t* and the two parental genotypes are *r s t* and *+ + +*. The possible recombinant products and their origins are:

NCO	r s t
NCO	+ + +
SCO in I	r + +
SCO in I	+ s t
SCO in II	r s +
SCO in II	+ + t
DCO	r + t
DCO	+ s +

Observation of the DCO classes above will show that, with respect to the allelic combinations of the two outside markers, the two alleles of the gene in the middle have been switched around: NCO are *r s t* and *+ + +*, and DCO are *r + t* and *+ s +*. While both NCO and DCO have the *r t* and *+ +* combinations, the two alleles of *s* have changed partners. This is what happens if *s* is the gene in the middle.

Returning to the data in the question, NCO are *pyr + met* and *+ ade +*, and DCO are *pyr ade met* and *+ + +*. Both have *pyr met* and the corresponding *+ +* combinations, but the two alleles of *ade* have been switched. Thus, we can determine from cursory observation which gene is in the middle, and therefore the order of the three genes.

Having done this, we only need to perform two calculations of genetic distance, those for interval I and interval II. The best estimate of the distance between the two outside markers is the sum of the two sub-intervals, as this calculation will include real crossovers that in DCO cancel each other out unless the middle marker is taken into account.

Using the estimates of crossing over in intervals I and II, the calculation of interference, as the coefficient of coincidence, proceeds as before.

Problem 18

Three-point cross in *Drosophila* (X-linked)

Question

In *Drosophila melanogaster*, a male from a strain true-breeding for three mutant characteristics, *w* (white eye), *m* (miniature wing) and *f* (forked thoracic bristles) is crossed to a female from a true-breeding wild-type strain. All the F_1 are phenotypically wild type. The F_2 females are similarly all phenotypically wild type, but the males have the following phenotypes and numbers:

white forked miniature	175
wild type	200
white forked	7
miniature	3
white	81
forked miniature	69
white miniature	42
forked	48

Interpret these data as fully as possible.

Answer

All three genes show different behaviour in the males and females of the F_2, as they segregate in the males but not in the females. All three therefore show the behavioural characteristics of sex-linked genes, genes located on the X chromosome (as shown in previous problems on sex-linkage). Therefore, all three genes are linked to each other.

The linkage of the three genes to each other is confirmed by the numbers of progeny in the various classes in the F_2 males. Here, because of the hemizygous condition in the XY males, the gametes from the female F_1 are reflected directly in the F_2 phenotypes, not masked as in their heterozygous sisters by the $+ + +$ sperm received from their F_1 fathers. On the assumption that eggs producing sons and eggs producing daughters were the same in types and frequencies, we can analyse this cross from the F_2 male data without the need for a test-cross.

As in the previous problem, we can identify the two largest (parental) classes, $w f m$ and $+ + +$, and the two smallest (DCO) classes, $w f +$ and $+ + m$. In this case, they differ by switching of the m gene between the two types, so m is the gene in the middle. The two largest classes are the parental ones, so we can now represent the original cross as:

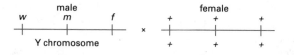

We now need to calculate the distance between genes w and m, and between m and f.

Considering genes w and m, the four possible combinations in the F_2 males and their numbers are:

w m $175 + 42 = 217$
w $+$ $81 + 7 = 88$
$+$ m $69 + 3 = 72$
$+$ $+$ $200 + 48 = 248$

Hence the distance between w and m is:

$$\frac{88 + 72}{217 + 88 + 72 + 248} \times 100\%$$

$$= \frac{160}{625} \times 100 = 25.6 \text{ map units}$$

Likewise for m to f:

m f $175 + 69 = 244$
m $+$ $42 + 3 = 45$
$+$ f $48 + 7 = 55$
$+$ $+$ $200 + 81 = 281$

The distance between m and f is:

$$\frac{45 + 55}{244 + 45 + 55 + 281} \times 100\%$$

$$= \frac{100}{625} \times 100 = 16.0 \text{ map units}$$

Hence the linkage map of the X chromosome is:

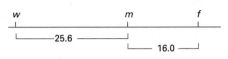

Finally, what is the coefficient of coincidence (extent and type of crossover interference) in this cross? From the lengths of the two intervals, we would expect 4.1% DCO (i.e. 25.6 in a sample of 625 F_2 males). We observe only 10 (7 + 3), giving us a coefficient of coincidence of about 0.39, showing quite strong positive interference.

| Problem 19 | Classic three-point cross in maize |

Question

In maize, *Zea mays*, plants grown from seeds of wild-type phenotype (coloured seed aleurone, starchy seed endosperm and plump seed endosperm) were test-crossed to a variety homozygous recessive for three genes, *c* (colourless aleurone), *w* (waxy endosperm) and *s* (shrunken endosperm).

In the progeny (seeds on the resulting maize cobs), the following were the seed phenotypic classes and numbers:

coloured, starchy, plump	17 959
colourless, starchy, plump	524
coloured, waxy, plump	4 455
coloured, starchy, shrunken	20
colourless, waxy, plump	12
colourless, starchy, shrunken	4 654
coloured, waxy, shrunken	509
colourless, waxy, shrunken	17 699

What was the genotype of the plant that was test-crossed? Derive a linkage map for the three linked genes segregating. Calculate the extent of any crossover interference.

The test-crossed plant was the result of an inter-cross between two inbred strains. Deduce their two genotypes.

Answer

Clearly the test-crossed plant was heterozygous for the three genes. As the test-cross is to a plant homozygous recessive for all genes under investigation, the test-cross progeny phenotypes directly reflect the gametes from the triple heterozygote. We can therefore use the table given in the question as the basis for the table of gamete types and frequencies shown below.

$+ + +$	17 959
$c + +$	524
$+ w +$	4 455
$+ + s$	20
$c w +$	12
$c + s$	4 654
$+ w s$	509
$c w s$	17 699

From the data provided, let us first deduce the genotypes of the parents of the test-crossed triple heterozygote, on the assumption that they were themselves true-breeding since we are told that they were both inbred lines. The two largest classes of gamete shown are $+ + +$ and $c w s$, so these result from a lack of any crossing over. They must therefore reflect the original parental gametes contributed to the test-crossed triple heterozygote. The genotypes must therefore have been $++ ++ ++$ and $cc\ ww\ ss$.

Now to determine the gene order, by the same criterion as in the worked answers to the preceding questions. The two largest classes (NCO) are $+ + +$ and $c w s$, and the two smallest (DCO) are $+ + s$ and $c w +$. The gene switched between the two parental combinations in the DCO combinations is the s gene, so this is between the other two. The gene order is therefore:

and the genotype of the triple heterozygote test-crossed is:

We can now refine the parental genotypes given above to indicate gene order also:

Continuing now with the calculation of inter-gene distances, the four allele combinations for $c–s$ are:

$c s$	$4654 + 17\,699 = 22\,353$
$c +$	$524 + 12 = 536$
$+ s$	$20 + 509 = 529$
$+ +$	$17\,959 + 4455 = 22\,414$

Therefore the distance between the c and s genes is:

$$\frac{536 + 529}{22\,353 + 536 + 529 + 22\,414} \times 100\% = \frac{106\,500}{45\,832} = 2.3 \text{ map units}$$

For genes s and w, the four allele combinations are:

$s\ w$ $509 + 17\,699 = 18\,208$

$s\ +$ $20 + 4654 = 4674$

$+\ w$ $4455 + 12 = 4467$

$+\ +$ $17\,959 + 524 = 18\,483$

The distance between genes s and w is therefore:

$$\frac{4674 + 4467}{18\,208 + 4674 + 4467 + 18\,483} \times 100\%$$

$$= \frac{914\,100}{45\,832} = 19.9 \text{ map units}$$

The full map of the three genes is therefore:

The expected frequency of DCO is 0.00458 (0.458% of the total progeny of 45 832), giving 209.9 DCO. The observed number of DCO is 32 (20 + 12), giving a coefficient of coincidence of 0.152, and hence quite strong positive crossover interference.

Problem 20 — Three-point cross in unordered tetrads

Question

In *Saccharomyces cerevisiae*, the following unordered tetrads were obtained from a sporulating diploid strain. The abbreviated gene symbols are *inl* (inositol requiring), *pdx* (pyridoxine requiring) and *thi* (thiamine requiring).

i
10 tetrads

inl	pdx	thi
inl	+	+
+	pdx	+
+	+	thi

ii
10 tetrads

inl	pdx	thi
inl	+	thi
+	pdx	+
+	+	+

iii
440 tetrads

inl	pdx	thi
inl	pdx	thi
+	+	+
+	+	+

iv
160 tetrads

inl	pdx	thi
inl	+	+
+	pdx	thi
+	+	+

v
10 tetrads

inl	+	+
inl	pdx	+
+	+	thi
+	pdx	thi

vi
10 tetrads

inl	pdx	+
inl	+	thi
+	pdx	thi
+	+	+

vii

360 tetrads

inl	pdx	thi
inl	pdx	+
+	+	thi
+	+	+

Interpret these data as fully as possible.

Answer

From these unordered tetrad data, we can investigate, as in previous tetrad analysis questions, the independent assortment or linkage of the genes and determine the genetic distances between any linked genes.

As before, we need to consider the genes a pair at a time, so let us start with the *inl* and *pdx* genes. Classifying the ascus types into the TT class and the two ditype classes, there are six TT classes, i, ii, iv, v and vi ($10 + 10 + 160 + 10 + 10 = 200$), and two ditype classes (iii and vii, $440 + 360 = 800$) which are both the same, and therefore presumably the PD category, i.e. the parental allelic combinations for the two genes under consideration. With 800 PD and 0 NPD, the two genes are clearly linked with the two NCO combinations *inl pdx* and *+ +*. Using the standard equation for calculating genetic distance from unordered tetrad data, the inter-gene distance is:

$$\frac{(200/2) + 0}{800 + 200} 100\%$$

$$= \frac{100}{1000} \times 100 = \frac{100}{10}$$

$$= 10 \text{ map units}$$

Moving on to *pdx* and *thi*, the TT are classes i, ii, v, vi and vii ($10 + 10 + 10 + 10 + 360 = 400$) and the ditypes, again all of the same type and hence the PD, are classes iii and iv ($440 + 160 = 600$). With no NPD, the two genes are clearly linked, and the distance between them is:

$$\frac{(400/2) + 0}{400 + 600} \times 100\%$$

$$= \frac{200}{1000} \times 100 = \frac{200}{10}$$

$$= 20 \text{ map units}$$

For *inl* and *thi*, the TT are classes i, iv, vi and vii ($10 + 160 + 10 + 360 = 540$) and there are two ditype classes, the larger of which are the PD:

inl	thi		inl	+
inl	thi		inl	+
+	+		+	thi
+	+		+	thi
classes ii and iii			class v	
$(10 + 440 = 450)$			10	

The inequality in the two ditype categories indicates linkage, and the distance between the genes is:

$$\frac{(540/2) + 10}{540 + 450 + 10} \times 100\%$$

$$= \frac{270 + 10}{1000} \times 100 = \frac{280}{10}$$

$$= 28 \text{ map units}$$

Hence the distance from *pdx* to *thi* is 20 map units, *inl* to *thi* 28 map units and *inl* to *pdx* 10 map units.

This indicates that the gene order is:

inl   pdx   thi
⊢— 10   —⊢————— 20 —————⊣

The full genotype of the triple heterozygote that was sporulated to give these asci was:

inl   pdx   thi
+   +   +

PROKARYOTIC GENETIC SYSTEMS

Introduction

The next set of problems move away from eukaryotes and their essentially Mendelian genetic systems to the bacteria and their viruses (bacteriophages, or *phages* for short). These show a number of distinctive features that complicate genetic analysis. This topic is treated in most comprehensive genetics textbooks, as well as in some specialized books. We will however outline here the chief characteristics of bacterial and phage genetics. More details will be found in the individual problems.

The main bacterial genome is a single DNA molecule that in almost all bacteria forms a closed loop, usually called the *circular chromosome*, otherwise found only in mitochondria and chloroplasts. Bacteria frequently also contain dispensable mini-chromosomes called *plasmids*. These may contain genes that confer extra properties on the host bacterial cell, e.g. resistance to various antibiotics (a matter of great medical importance), production of agents that kill competing bacteria, ability to exploit additional environmental niches, or pathogenicity for animals (including humans) or plants. Very often these genes are carried within transposable elements, of the type termed *transposons*.

Phage genomes, like those of other viruses, can be of all kinds: DNA or RNA, single- or double-stranded, linear or circular. Phages typically show a lytic cycle that corresponds to much the same life cycle as other viruses: the phage particle infects a sensitive host cell, the phage nucleic acid replicates intracellularly, phage coat components are synthesized, phage particles are assembled, and eventually the cell lyses with release of the phage.

Plasmids are not the only kind of additional genetic elements that bacteria may contain. Some phages are said to be *temperate*, meaning that their DNA can be maintained within the bacterial cell in a quiescent state (termed *prophage*). This quiescent state comes about because the phage encodes a repressor that switches off expression of genes encoding lytic functions. A bacterium carrying a prophage is said to be *lysogenic* for that phage (it is a lysogen and has been lysogenized). A phage that cannot switch off lytic functions is termed *virulent*. The prophage is most often integrated into the host chromosome, but sometimes takes the form of a plasmid. The quiescent state of the prophage can give way to renewed expression of the lytic functions, a process termed *induction*. When the prophage is integrated into the chromosome, induction also requires *excision*, the reverse of the integration step. Temperate phages sometimes possess genes of the same kinds as plasmids that confer useful properties on their host cells; and as with plasmids, such genes often lie within transposons. Bacteria, like higher organisms, may carry transposable elements, lengths of DNA that can move from one location to another (which in this case could be in the main chromosome, a plasmid or a phage).

DNA transfer   Bacteria do not have any equivalent of the eukaryote zygote, which carries a complete set of genetic information from both parents. Instead, there are various forms of DNA transfer in which DNA – almost always a fragment and often only a small one – of the total genome of a 'donor' cell is transferred into a 'recipient' cell, which contributes its entire genome. (We still use the term 'cross', though.) A complete circular plasmid can be transferred and may be able to perpetuate itself in the recipient. However, chromosomal DNA will necessarily be a linear fragment and it is this that gives bacterial genetics its peculiar (and especially difficult!) features. A single crossover between the incoming linear fragment and the resident circular chromosome gives rise to a linear chromosome with repeats at the ends, something which cannot replicate. For this reason, only double crossovers (or, more generally, an even number) are possible.

DNA transfer from donor to recipient can occur by three distinct mechanisms: transformation, transduction and conjugation. In *transformation*, DNA is released from donor cells and taken up as such by the recipient. In *transduction*, donor DNA finds its way into a phage particle as the latter develops inside an infected host cell, remains present within the phage when the latter is released on lysis of the host, and is transferred into a recipient cell when the latter is infected by the phage particle. *Conjugation* is mediated by cell-to-cell contact; it seems always to require the presence in the donor cell of a plasmid encoding the necessary functions. Such a plasmid is termed *conjugative*; plasmids that do not encode conjugation functions are thus non-conjugative.

A minority of conjugative plasmids can integrate into the chromosome, usually at a variety of sites (though not completely at random). The most studied of these is the *Escherichia coli* F plasmid (you may find this called the 'F factor' in older texts). Cells in which F is chromosomal are termed 'Hfr', for high frequency of recombination, because they donate chromosomal DNA fragments more often than cells (termed 'F$^+$') in which F is in the plasmid state. (A cell lacking F completely is termed 'F$^-$'.) Plasmids, like F, that can integrate into the chromosome have this ability in common with one kind of temperate phage (see above) of which the best-known example is the *E. coli* phage λ. To cover their shared ability, the term *episome* was coined some years ago; an episome is therefore any genetic element that can either replicate autonomously or integrate into the host chromosome. There are some problems with this term, however, and it is now less often used.

Bacterial genetic   Genetic analysis in bacterial crosses depends on the nature of the chromosomal
analysis   fragments. In transformation (Problem 27), the fragments are small and their ends can be anywhere within the chromosomal DNA (*random ends*). As for transduction, it turns out that there are two kinds, *generalized* and *specialized*. In generalized transduction, the fragments have much the same features as in transformation and since it does not introduce any new feature of genetic analysis, there is no problem here that is based on it. Specialized transduction (Problem 26) is mediated only by those temperate phages which, like λ, integrate into the chromosome: the fragment (again small) is always of DNA that lies adjacent to the integrated prophage. Specialized transduction comes about in rare instances where excision goes slightly

wrong (*aberrant excision*) and, instead of producing the normal phage genome, results in the formation of a genome in which (usually) a part of the phage genome is replaced by a roughly equal amount of host chromosomal DNA. Both integration and excision involve a type of recombination; the incorrect sort involved in aberrant excision is termed *illegitimate recombination*.

Chromosomal DNA transfer by conjugation is liable to result in the transfer of much larger pieces of DNA than with transformation or transduction, even (with very low frequency) of the entire donor chromosome. Sometimes the fragments are again random; however, much the best researched system is that for *E. coli* and the F plasmid, in particular where the donor is an Hfr cell, i.e. F is integrated. Although this is often taken as *the* model of conjugational DNA transfer, it has many unrepresentative features. Nevertheless, we will concentrate on it here because it illustrates all possibilities. When F is present as an autonomous plasmid, it normally promotes conjugation purely for its own transfer, like any other conjugative plasmid. In an Hfr, F again tries to promote its own transfer, but the result now is that pieces of chromosomal DNA are transferred one of whose ends is within F. Hence, unlike the situation in transformation and generalized transduction, one end of the transferred fragment is not random.

A remarkable feature of the Hfr × F⁻ cross is that transfer of the chromosomal fragment is surprisingly slow: it takes 90 minutes for passage of the entire donor chromosome. In addition, the mating bridge between donor and recipient is fragile enough to be broken, by violent agitation for example. This permits the construction of a physical map by *interrupted mating* (Problems 21 and 23), as well as the use of standard recombinational methodology (Problem 22).

Like λ, F can excise from its chromosomally integrated state; and again like λ, excision can be aberrant through illegitimate recombination. The result is a modified F plasmid that carries a piece of chromosomal DNA that lay adjacent to it when it was integrated in the Hfr cell. Such a modified F is termed an 'F′', read as 'F-prime'. An F′ may transfer via conjugation as efficiently as the original F. This system is the subject of Problem 24.

Phage genetics

After the oddities of bacterial genetic analysis, phage genetics comes as a relief. Phage crosses simply involve co-infection of a host with genetically marked variants of the same phage. Recombination therefore occurs between complete linear genomes and analysis thus resembles ordinary haploid genetics (Problem 28).

Problem 21

Interrupted mating using time of transfer to map a part of the *E. coli* chromosome

Question

The transposon Tn*5*, which confers resistance to kanamycin, inserts randomly into *E. coli* DNA. As part of a project to obtain insertions of Tn*5* into many different sites in the *E. coli* chromosome, a scientist obtained a Tn*5* insertion in the Hfr strain Hfr1,

Time of sampling (min)	Medium A	Medium B	Medium C
0	0	0	0
3	0	0	0
6	0	0	0
9	0	14	0
12	13	75	0
15	49	165	0
18	100	198	6
21	139	204	23
24	160	206	51
27	171	208	80
30	175	209	99
33	175	209	108

Table 21.1. Numbers of recombinant colonies grown on different media and measured at different sampling times.

which is prototrophic and sensitive to antibiotics. She then proceeded to map this insertion by interrupted mating. Preliminary tests suggested that in the insertion strain (written Hfr1::Tn5), Tn5 had inserted near a region of the chromosome that includes a gene cluster *gal* for utilization of the sugar galactose as sole carbon source, and another cluster *trp* for synthesis of the amino acid tryptophan. The strategy was to mate Hfr1::Tn5 with an F⁻ strain called X100, that was *galK⁻ trpA⁻ rpsL* but had no other markers (*rpsL* indicates a mutation that confers resistance to streptomycin; it affects one of the proteins of the small ribosomal subunit). Prototrophic *E. coli* strains grow on minimal salts medium with glucose as sole carbon source: this will be written Min + glu. Hfr1::Tn5 grows on Min + glu or Min + gal (the latter contains galactose instead of glucose), and also on Min + glu containing kanamycin (Min + glu + kan) but not on Min + glu containing streptomycin (Min + glu + str). As expected, X100 grows on Min + glu supplemented with tryptophan (Min + glu + trp) but not on Min + gal + trp or on Min + glu; it also grows on Min + glu + trp + str but not on Min + glu + trp + kan. An exponential culture of Hfr1::Tn5 and a stationary phase culture of X100 were grown up. These were mixed to give 10^6/ml of the former and 10^8/ml of the latter, and the mixture incubated at 37 °C. At 3-minute intervals from the time of mixing, samples were taken and instantly subjected to violent mechanical agitation, diluted 100-fold and 0.1 ml aliquots plated on the following media (they are given letters for ease of reference later): (A) Min + glu + trp + kan + str; (B) Min + gal + trp + str; and (C) Min + glu + str. The numbers of resulting colonies, for different times of sampling, are shown in Table 21.1.

From the above data:

1. Using graph paper, plot the number of colonies on each medium against time of sampling.
2. What genotypes are selected on each medium? How do they arise?
3. Why is the final number of colonies different for the three media?
4. Why is the slope of the linear part of each plot different?
5. What is the map position of each marker mapped in this experiment? (Make clear what units you are using to define map position.)

Answer

1. Figure 21.1 shows a typical curve. Each plot is sigmoid, with a more-or-less linear central segment. (The reason why this approximates to linearity is mathematically not obvious; the only justification at this level is empirical, though mathematical calculations exist.) The slope is steepest for medium B and shallowest for medium C. Similarly, the value at which the number of colonies levels off is greatest for medium B and least for medium C.

2. Min + glu + trp + kan + str selects for cells that combine the kanamycin resistance (*kan*^{r}) of the Hfr with the streptomycin resistance (*str*^{r}) of the F$^-$; Min + gal + trp + str selects for cells that combine the *gal*$^+$ character of the Hfr with the *str*^{r} of the F$^-$; and Min + glu + str selects for cells that combine the *trp*$^+$ character of the Hfr with the *str*^{r} of the F$^-$. Each medium, by selecting for cells with some properties from both parents, selects for *recombinants*. These three media select for the recombinant *genotypes kan*^{r} *str*^{r}, *gal*$^+$ *str*^{r} and *trp*$^+$ *str*^{r} respectively. (This can be expressed slightly differently, and perhaps more correctly, by saying that these media select for the *phenotypes* Kan^r Str^r, Gal$^+$ Str^r and Trp$^+$ Str^r. The convention is that roman letters/capital denote phenotype, italic/lower case denote genotype. In a simple case like this, the two are virtually interchangeable, but this is not always so.) How these recombinant genotypes are generated will now be considered.

3. This relates to the mechanism by which a recombinant is generated from an Hfr × F$^-$ mating pair of cells. A single strand of Hfr DNA, beginning at the transfer origin of replication of the inserted F, is threaded into the F$^-$. The Hfr cell is thus the *donor* of DNA to the *recipient* F$^-$. Even if the mating is uninterrupted, there is a constant probability per unit time that the mating pair will break apart, leaving the linear fragment of Hfr DNA within the F$^-$. If, as here, the mating is

Figure 21.1. Plots of number of recombinant colonies against time of sampling, using data from Table 21.1.

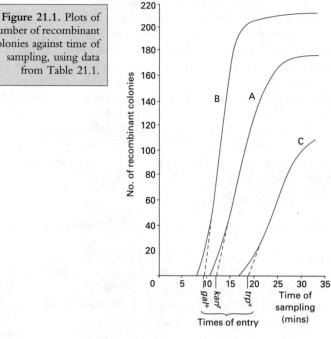

deliberately *interrupted*, all mating pairs in the mixture separate at the time of interruption. Consider an Hfr marker M that begins to enter the F^- x minutes after the initiation of mating. Many mating pairs in the mixture will have separated before the x minutes have elapsed. As DNA transfer is not initiated synchronously in all mating pairs, M will be transferred in different mating pairs for several minutes after x (this is why the plots in (1) slope, rather than going vertically upwards). After some time, M will have been transferred in all the mating pairs still extant, so that the plot reaches a plateau. The greater x is, the higher the proportion of mating pairs originally formed that will have separated before x, and thus the lower the plateau value. See also Problem 22.

4. This is not obvious either. Not only is DNA transfer not initiated synchronously in all mating pairs, it does not proceed at a uniform rate. The longer mating proceeds, the greater will be the variation between mating pairs in the length of DNA transferred. Hence the later a marker is transferred, the longer the interval between its first and last entries into the F^-, i.e. the lower the slope of the plot.

5. The time of entry is empirically estimated by extending the linear part of the plot back (shown by the dashed line in Fig. 21.1) to cut the time axis; the intercept is taken as the time of entry. In this case, the times of entry of the *gal*$^+$, *kan*^{r} and *trp*$^+$ markers from the Hfr are about 9.5, 12 and 18.5 minutes respectively. These times of entry are *map positions*, which are in time units (minutes).

Problem 22

Conjugation with back-marker mapping

Question

This follows on from Problem 21. Another cross was performed, using the same Hfr1::Tn*5* together with a derivative of X100 termed X101. X101 has the same *galK*$^-$ *trpA*$^-$ *rpsL* markers as X100, but has three additional ones: *argG*, *hisD* and *metA*, which confer additional nutritional requirements for the amino acids arginine, histidine and methionine respectively. The Hfr1::Tn*5* × X101 cross was carried out exactly as in Problem 21, but the mating was *uninterrupted*: the mating mixture was incubated undisturbed at 37 °C for 2 hours prior to sampling, diluting and plating as before. The selective plates used, and the numbers of colonies found on each, were as follows (arg, his and met stand for arginine, histidine and methionine):

(D) Min + glu + trp + arg + his + met + kan + str, 172
(E) Min + gal + trp + arg + his + met + str, 206
(F) Min + glu + arg + his + met + str, 110
(G) Min + glu + trp + his + met + str, 5
(H) Min + glu + trp + arg + met + str, 18
(I) Min + glu + trp + arg + his + str, 1

1. Using semi-log graph paper, plot the percentage of Hfr cells that have given rise to colonies on media D, E and F against the map positions of the markers as deduced in Problem 21. Draw a line through the three points and read off the rough map positions corresponding to the markers selected on media G, H and I.

Table 22.1. Percentage of colonies from each medium when grown on every other medium.

Colonies taken from	Percentage that grew when tested on:					
	D	E	F	G	H	I
D		80	64	3	10	1
E	83		53	2	9	0
F	75	60		5	16	1
G	50	50	50		53	20
H	53	52	55	28		50
I	50	50	50	54	50	

Table 22.2. Response of colonies from medium F when grown on media D and E.

Growth on D and E	50%
Growth on D but not E	20%
Growth on E but not D	4%
Growth on neither D nor E	21%

2. Colonies from each medium were tested on all the other media. The percentages that grew are shown in Table 22.1. Explain these figures. You might like to consider in particular: why are so many at or just above 50%? Why are some getting on for 100%, while others drop well below 50%?

3. In addition, the individual response of each colony from medium F when grown on media D and E was noted and the results are shown in Table 22.2. Explain these figures too.

Answer

1. Note that the genotypes (see Problem 21, question 2) selected by the various media are:

   D, kan^r str^r
   E, gal^+ str^r
   F, trp^+ str^r
   G, arg^+ str^r
   H, his^+ str^r
   I, met^+ str^r

From the information given in Problem 21, each 0.1 ml diluted mating mixture plated contains 10^3 Hfr cells. Hence the 172 colonies on medium D represent a percentage of the Hfr cells of $(172/10^3) \times 100 = 17.2\%$; and so on for the other media. Using semi-log paper, one can plot the three percentages for media D, E and F against times of entry from Problem 21. The points fall roughly on a straight line, though with only three points that does not mean much! This line can be extended and the times of entry for the three markers of unknown location arg^+, his^+ and met^+ can be read off using the calculated percentages. They are approximately 68, 44 and 89 minutes respectively.

Table 22.3. Data from Table 22.1 showing whether second marker is transferred before or after the first.

Colonies taken from	Tested on					
	D	E	F	G	H	I
D		b	a	a	a	a
E	a		a	a	a	a
F	b	b		a	a	a
G	b	b	b		b	a
H	b	b	b	a		a
I	b	b	b	b	b	

2. These percentages are a measure of genetic *linkage*, i.e. frequency of co-inheritance in recombinants, under the conditions of the experiment. When a recombinant selected on one medium is tested on another selective medium (Table 22.1), the second marker tested ('grew when tested on') must be transferred either *before* or *after* the first ('colonies taken from'). By the end of question (1) you will have approximate times-of-entry map positions for all six markers. Therefore you can see, for each combination in Table 22.1, whether the second marker is transferred before or after the first. This is shown in Table 22.3 where *a* indicates second marker transferred after first and *b* indicates second marker transferred before first. The first thing to notice here is that whereas *a* linkage values vary from 83% to zero, *b* linkage values never go below 50%. The explanation for *a* is because of spontaneous separation of mating pairs; in many cases where the selected earlier marker has been transferred, the later marker will never be transferred because the mating pair has broken up before the second marker is reached. The percentages are in fact directly related to the numbers of colonies found on the various selective media, e.g. for medium D this is 172, for medium H 18, and the figure for 'taken from D/grew when tested on H' is (172/18) ×100 = 10%. The explanation for *b* is as follows. For an earlier marker transferred only shortly before the later selected one, it is unlikely that any crossover will occur between the two; hence the linkage will be high, approaching 100% for very closely located markers. For an earlier marker transferred long before the selected one, there will be many crossovers between them. If the number of crossovers is even, the markers will be inherited together by the recombinant; if the number of crossovers is odd, the earlier marker will not be inherited. There is a 50% chance of the number of crossovers being either even or odd; hence the 50% figure for the linkage between the last selected markers and the earliest markers to be transferred.

3. Figure 22.1 shows the three markers in their arrangement in the Hfr and F⁻ parents. The regions to the left of *galK*, between *galK* and Tn5, between Tn5 and *trpA*, and to the right of *trpA* are denoted I, II, III and IV respectively. Medium F selects for colonies growing from cells with the recombinant genotype trp^+ str^r; these must have one of the four possible genetic combinations of the *galK* and Tn5 markers, namely gal^+ Tn5⁺, gal^+ Tn5⁻, gal^- Tn5⁺ and gal^- Tn5⁻ (Tn5⁺ and Tn5⁻ are merely shorthand for 'Tn5 present' and 'Tn5 absent' respectively). In Table 22.2, 'Growth on D and E' corresponds to genotype gal^+ Tn5⁺, this

Figure 22.1. Cross-over patterns that generate the various recombinant classes in Problem 22, part 3. Parts (a) to (d) show the crossover positions that generate the four possible *gal*⁺ Tn*5* recombinant types.

Possible regions where crossing-over can occur:

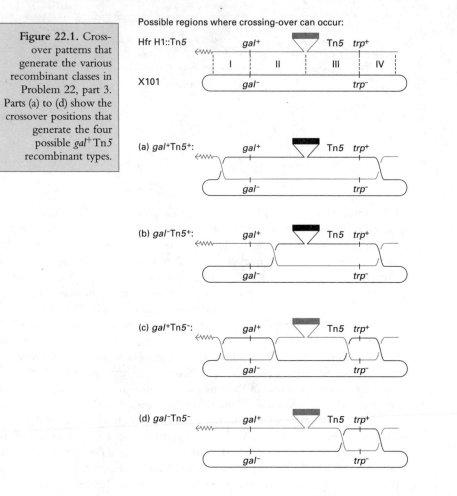

recombinant class resulting from crossovers in regions I and IV (Fig. 22.1a); 'Growth on D but not E' corresponds to genotype *gal*⁻ Tn*5*⁺, this recombinant class resulting from crossovers in regions II and IV (Fig. 22.1b); 'Growth on E but not D' corresponds to genotype *gal*⁺ Tn*5*⁻, this recombinant class resulting from crossovers in regions I, II, III and IV (Fig. 22.1c); and 'Growth on neither D nor E' corresponds to genotype *gal*⁻ Tn*5*⁻, this recombinant class resulting from crossovers in regions III and IV (Fig. 22.1d). Note that the rarest class is the only one requiring four, rather than two, crossovers. If we had not already known (from the interrupted mating experiment in Problem 21) that Tn*5* was the middle marker of the three, it would have been possible to deduce as much from these results alone. This criterion, that *the rarest recombinant class results from the most complex crossover pattern*, is used a great deal to order markers in both prokaryote and eukaryote genetics. See also Problems 26 and 27. (Actual percentages can also be correlated with linkage data from question (2).)

Conjugation mapping of the *E. coli* chromosome using different Hfr strains to generate a circular map

Question

Two further interrupted mating experiments were carried out in the same way as in Problem 21, except that samples were taken at 5- rather than 3-minute intervals. X101 was again used as recipient, while the Hfr donors were two uncharacterized ones termed Hfr2 and Hfr3, again prototrophic and sensitive to streptomycin. Media E, F, G, and H as in Problem 22 were used for selection: in the cross Hfr2 × X101, media E, F and H were used; in the cross Hfr3 × X101, media F, G and H. Numbers of recombinant colonies obtained on the various selective media at different sampling times are shown in Tables 23.1 and 23.2.

1. Calculate the map position, as time of entry, of each of the selected markers in the two crosses as in Problem 21.
2. Explain the apparent map positions of the markers in these crosses in comparison with those deduced in Problem 21.

Table 23.1. Numbers of recombinant colonies obtained on various media at different sampling times from cross Hfr2 × X101.

Time of sampling (min)	Medium E	Medium F	Medium H
0	0	0	0
5	0	0	33
10	0	0	68
15	0	0	118
20	0	8	157
25	0	26	187
30	4	48	209
35	14	73	224
40	26	97	236
45	40	118	243
50	53	134	248
55	63	145	252

Table 23.2. Numbers of recombinant colonies obtained for various media at different sampling times from cross Hfr3 × X101.

Time of sampling (min)	Medium F	Medium G	Medium H
0	0	0	0
5	14	0	0
10	109	0	0
15	222	0	0
20	328	0	0
25	357	0	33
30	360	0	78
35	353	0	127
40	367	0	186
45	359	3	218
50	362	20	242
55	355	41	250

Answer

1. The times of entry (in minutes) are approximately (depending on just how you draw the best line): Hfr2 × X101, his^+ 4, trp^+ 21, gal^+ 31; Hfr×X101, trp^+ 5, his^+ 22, arg^+ 46.

2. The map positions can be reconciled by assuming that: (i) the *E. coli* genetic map is 'circular'; (ii) the F plasmid, whose change from the autonomous (plasmid) to the integrated state is the basis for the change of an F^+ to an Hfr strain, can integrate at many different sites and in either orientation; and (iii) the point of origin (i.e. the first point to be transferred) and orientation of transfer (i.e. clockwise or anticlockwise) of an Hfr reflect the location of the inserted F and its orientation, which may differ as between different Hfr strains. So the origin of Hfr1 (from Problem 21) is 9.5 minutes anticlockwise of *gal* and it transfers clockwise; the origin of Hfr2 is 4 minutes clockwise of *his* and it transfers anticlockwise; and the origin of Hfr3 is 5 minutes anticlockwise of *trp* and it transfers clockwise (Fig. 23.1).

Figure 23.1. Positions of markers and Hfr origins of transfer from interrupted mating experiments in Problems 21 and 23. The figures inside the circle that depicts the *E. coli* chromosome are map positions inferred from marker times of entry.

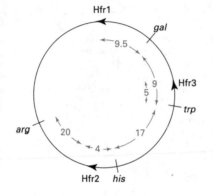

Problem 24

Properties of F′ strains

Question

Hfr2, described in Problem 23, was used in a further cross as follows. The recipient employed was X102, which carried the same *hisD* and *rpsL* markers as X101 (but none of the others), together with an additional marker *purF⁻*. Strains carrying *purF⁻* require supplementation of minimal salts plus glucose medium with the purine base adenine (abbreviated below to ade). Whereas *hisD* is transferred very early by Hfr2 (see Problem 23), *purF* was found to be transferred very late; this indicates that in Hfr2, F had integrated between these two genes. Hfr2 was crossed with X102 in an interrupted mating under the standard conditions described in Problem 21; however, only one sample was taken, in which mating was interrupted after 60 minutes. Transfer of the entire *E. coli* chromosome requires at least 90 minutes. The mating mixture, and separate parental controls, were plated on (A) Min + glu + his + str, and (B) Min + glu + ade + str.

The parental controls gave no colonies; the mating mixture sample gave three colonies on (A) and some thousands of colonies on (B). The three colonies on (A), and a representative colony on (B), were patched on to the same medium to give stocks of the four isolates, which were termed X, Y, Z and W. These isolates were tested further as follows.

(a) X, Y and Z were tested on medium (B) and W on medium (A). X and Z grew on (B), while Y did not; W did not grow on (A).

(b) The next experiment was designed to test the ability of X, Y, Z and W to act as donor in transferring various markers to a recipient strain. Because these isolates all carry the *rpsL* marker used in the Hfr2 × X102 cross to select against the donor Hfr2, a new marker had to be employed capable of selecting against X102-derived donors. Accordingly, a derivative called X103 was obtained from X102 that carried: (i) all the markers of X102; and (ii) a new marker *gyrA*, which confers resistance to the antibiotic nalidixic acid (nal). X, Y, Z and W were crossed with X103 in uninterrupted matings, as in Problem 22, and plated (undiluted and at various dilutions) on the following media: (C) Min + glu + his + nal and (D) Min + glu + ade + nal. Numbers of colonies, corrected where necessary for dilution, for the various classes of isolate are shown below.

Medium	X, Z	Y	W
(C)	10 000	10 000	0
(D)	10 000	0	0

1. Bearing in mind that Hfr2 transfers *purF* as a terminal marker and that transfer of the complete *E. coli* chromosome takes 90 minutes, how could colonies arise on medium (A) after only 60 minutes of mating?
2. Explain the pattern of further transfer of markers by X, Y and Z, and non-transfer of markers by W, as shown above.

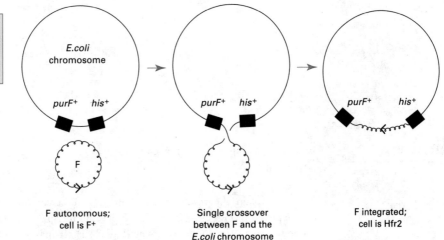

Figure 24.1. Generation of F's: integration of F to give Hfr2.

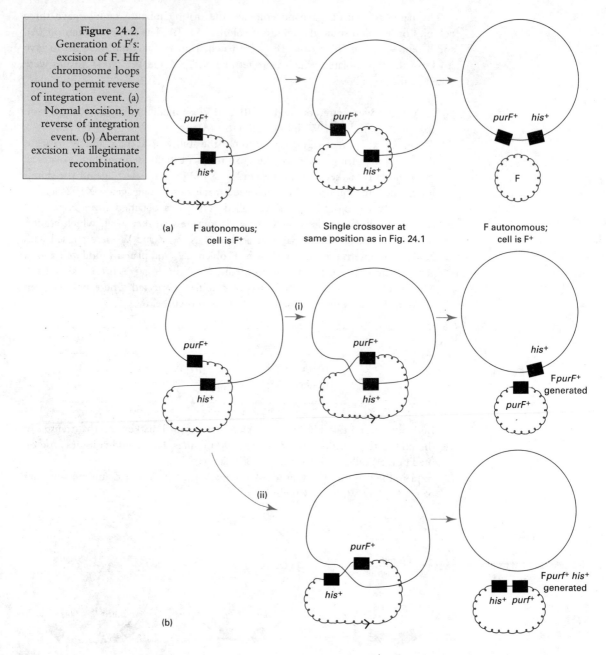

Figure 24.2. Generation of F's: excision of F. Hfr chromosome loops round to permit reverse of integration event. (a) Normal excision, by reverse of integration event. (b) Aberrant excision via illegitimate recombination.

(a)   F autonomous; cell is F+

Single crossover at same position as in Fig. 24.1

F autonomous; cell is F+

(b)

Answer

Medium (A) selects for Ade$^+$ Str^r, (B) for His$^+$ Str^r, (C) for Ade$^+$ Nal^r and (D) for His$^+$ Nal^r.

1. To explain the results detailed in (a), we have to invoke the genetic events depicted in Figs 24.1–24.5. The F plasmid integrates into the *E. coli* chromosome by means of a single crossover (Fig. 24.1). It may *excise* in the same way, by a

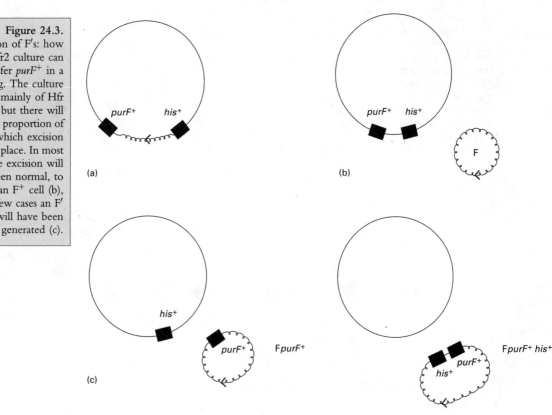

Figure 24.3.
Generation of F's: how
an Hfr2 culture can
transfer *purF*$^+$ in a
mating. The culture
consists mainly of Hfr
cells (a), but there will
be a small proportion of
cells in which excision
has taken place. In most
cases, the excision will
have been normal, to
generate an F$^+$ cell (b),
but in a few cases an F′
will have been
generated (c).

reverse of the original event (Fig. 24.2(a)). *Aberrant excision* may however also occur at low frequency, by *illegitimate recombination* between sites that perhaps share only a limited degree of fortuitous homology. If one of these sites lies in chromosomal DNA on the 'late' side of the integrated F, then a modified version F, called F′, will result that carries a piece of chromosomal DNA which would normally be transferred very late; in this case, it contains the Hfr2 *purF*$^+$ marker (Fig. 24.2(b)). If such an F′ arises, it will normally be small by comparison with the entire chromosome and thus transfer quickly. Hence X, Y and Z will each contain an F′ carrying *purF*$^+$; these can be written F*purF*$^+$. The other site of illegitimate recombination may lie within F or in chromosomal DNA on the 'early' side of the integrated F. Hence some, but not all, F*purF*$^+$ may also carry *his*$^+$ from Hfr2; those that do may be termed F*purF*$^+$ *his*$^+$ (Fig. 24.2(b)ii). Some of the colonies growing on (A) could be of this type, though others may have acquired an F*purF*$^+$ that lacks *his*$^+$ (Fig. 24.2(b)i) (note that it would not be correct to term this an F*purF*$^+$ *his*, which would suggest the presence on the F′ of one or more defective *his* genes).

2. The experiment described in (b) shows that X, Y and Z are capable of further high-frequency conjugal transfer of markers. X and Z transfer both *purF*$^+$ and *his*$^+$

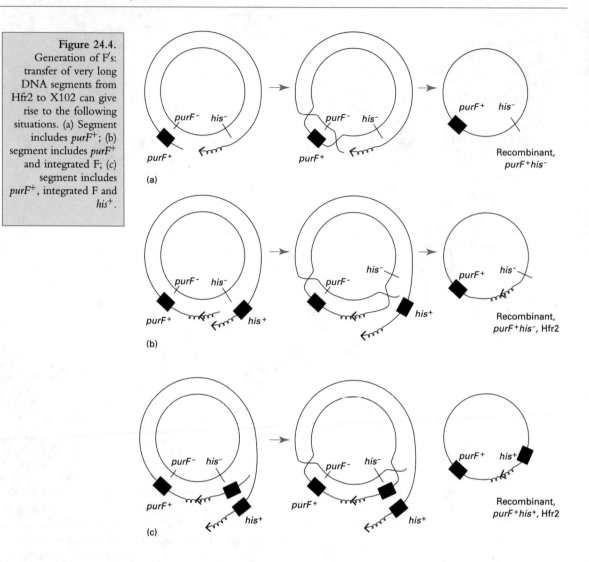

Figure 24.4. Generation of F's: transfer of very long DNA segments from Hfr2 to X102 can give rise to the following situations. (a) Segment includes *purF+*; (b) segment includes *purF+* and integrated F; (c) segment includes *purF+*, integrated F and *his+*.

to X103, while Y transfers only *purF+*. We may conclude that X and Z acquired an F*purF+ his+* and Y an F*purF+*.

3. Isolate W is an ordinary recombinant for a marker transferred proximally by Hfr2. It will therefore not carry F at all (it will be F−) and cannot transfer its *his+* marker to X103 by conjugation.

The products of F′ transfer are *partial diploids*, i.e. they contain two copies of the chromosomal segment carried on the F′, the other of course being within the chromosome (Fig. 24.3). In this respect, they resemble *specialized transductants* (Problem 25). Like the latter, they can be used in *complementation testing*.

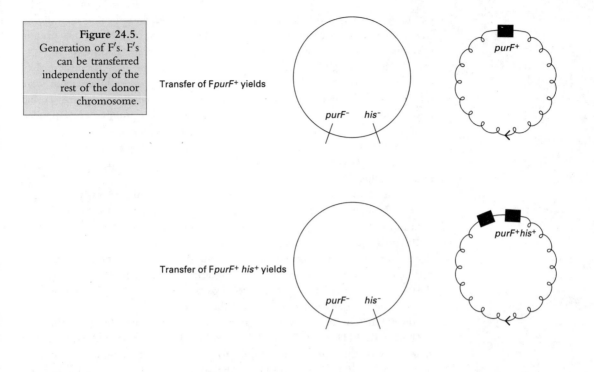

Figure 24.5.
Generation of F's. F's can be transferred independently of the rest of the donor chromosome.

Transfer of F*purF⁺* yields

Transfer of F*purF⁺ his⁺* yields

Problem 25 | Temperate phages and lysogenization

Question

The *Salmonella typhimurium* phage P22 is a conventional temperate phage. Like λ, its prophage is integrated into the host chromosome (though you do not need this fact to solve the following problem). Also characteristically of a temperate phage, it forms *turbid* plaques, whereas most (though not all) virulent phage/host combinations show *clear* plaques. (The reason for using P22 here rather than λ is that λ plaques are small and rather indistinct; their turbidity is not very clear, so to speak! P22 plaques are big and obviously turbid.) This problem, which is descriptive rather than quantitative, relates to the difference between clear and turbid plaques, and the nature of the turbidity.

Serial dilutions of a suspension of P22 phage particles were plated in soft agar (on a nutrient agar base, as is conventional in phage work) with a culture of a sensitive *S. typhimurium* host not lysogenic for P22. On a plate with about 50 plaques, the plaques were large and turbid. A plate with a 100-fold lower dilution of P22 particles showed almost all the 5000 or so plaques to be much smaller because of crowding, but still turbid. However, a few plaques on the latter plate were visibly different in being clear. The following experiments were then carried out. Explain the results.

1. Bacterial growth from within a turbid plaque was transferred to a fresh plate. Two colonies that resulted, termed ST1 and ST2, were grown up in liquid broth and tested as follows.

(a) They were plated in soft agar with a P22 suspension, just like the *S. typhimurium* culture used at the start of the problem. However, no plaques developed, even when as many as 10^8 phage particles were used per plate.

(b) The ST1 and ST2 cultures were centrifuged and the supernatants freed of traces of bacteria by membrane filtration. Serial dilutions of the supernatants were then plated with the P22-sensitive, non-lysogenic *S. typhimurium* culture. Turbid plaques, looking like typical P22 plaques, were seen; their number corresponded to about 10^7 plaque-forming units per ml of culture supernatant. ·

2. Some of the clear plaques from among the 5000 turbid plaques were purified and phage stocks grown up from them. Three of these, PH1, PH2 and PH3, were tested as follows.

(a) Serial dilutions were plated with the P22-sensitive, non-lysogenic *S. typhimurium* culture; clear plaques were seen, as expected. They were also plated with isolates ST1 and ST2; this time no plaques were found, even when as many as 10^8 phage particles were used per plate.

(b) A soft agar layer containing only the P22-sensitive, non-lysogenic *S. typhimurium* cells was poured on a nutrient agar base and allowed to set. Drops of PH1, PH2 and PH3 phage suspensions were then placed on the soft agar, as follows: each suspension alone, and also the wild-type P22 as a control; and pairs of drops that were allowed to mix, in the three possible combinations PH1 + PH2, PH1 + PH3 and PH2 + PH3. The wild-type P22 gave a turbid zone of lysis (this corresponds simply to the superposition of lots of separate plaques), while PH1, PH2 and PH3 on their own gave clear zones, much as might have been expected. As for the mixtures, PH1 + PH2 and PH2 + PH3 gave turbid zones, while PH1 + PH3 gave a clear zone.

Answer

1. What is the nature of the growth within the turbid plaque? Remember that a temperate phage like P22 can, after infection, either *lyse* or *lysogenize* the infected cell. A plaque starts from a single infected cell; when this lyses, the released phages swim out a little way through the soft agar and infect a new set of cells; the process continues like a chain reaction until all growth on the plate ceases through nutrient depletion or accumulation of toxic products, and a visible plaque is the result. What happens in those cases where an infected cell lysogenizes rather than lyses? It will still be infected by new phage particles: in the case of P22 and λ, and most though not all temperate phages, a lysogen can still be infected. But the prophage within the lysogen carries a gene encoding a *repressor* of lytic functions, and this gene is expressed in the lysogen (at least, until induction occurs). The repressor ensures that the prophage's own lytic functions are not expressed, *but it will do the same for any invading phage genome of the same type*. Hence the lysogen is *immune to superinfection* by the same kind of phage; and so lysogens give rise to tiny colonies within the plaque, rendering the latter turbid.

Cells transferred from within the turbid region, such as ST1 and ST2 here, are therefore lysogens. Test (a) is explained by their superinfection immunity, which means that the wild-type P22 cannot form plaques on them. Test (b) reflects the phenomenon of induction – the escape of the prophage from quiescence. This occurs spontaneously and at random at a low frequency, and results in a small fraction of the lysogenic population constantly lysing and giving rise to phage particles. These remain in the culture medium and end up in the supernatant when the culture is centrifuged. They then give rise to plaques when plated with the non-lysogenic host.

2. What causes the clear plaques? If turbidity in a plaque is due to lysogenization, a clear plaque must result because the phage has lost the capacity to lysogenize, presumably through mutation. In fact, clear-plaque (*c*) mutants are characteristic of temperate phages. The commonest defect is inability to make repressor. However, *c* mutants usually retain the capacity to respond to repressor by switching off lytic functions. So the explanation of test (a) is that when the *c* mutant genome of isolates PH1, PH2 or PH3 enters a lysogen such as ST1 or ST2, the repressor produced by the prophage represses lytic functions in the incoming genome as well as in the prophage, and no plaques are formed.

   Test (b) introduces the *complementation test*. The simplest explanation of the results with the phage mixtures is as follows. There may be two or more genes in the P22 genome whose products are required for the formation of repressor: for instance, one gene may encode the repressor protein and another gene may be needed for its production to get going after infection. Suppose that PH1 is mutated in one of these genes (call it *c1*) and PH2 in a different *c* gene *c2*. PH1 is therefore necessarily $c2^+$ and PH2 is $c1^+$. So when the two mutant genomes co-infect the same cell, a situation that will occur in the region of the mixture of phage suspensions in this experiment, PH1 directs the formation of the product of *c2* and PH2 that of the product of *c1*. In other words, in the mixed infected cells, both the gene products necessary for lysogenization are made and the cells may be lysogenized with the formation of a turbid zone. Suppose now that PH3 is also mutated in *c1*. Co-infection by PH1 and PH3 leaves no possibility of functional *c1* product being made and mixed infected cells can never lysogenize, so the zone remains clear. This exemplifies the basis of the complementation test. The aim of the test is to find out whether two mutations, present in separate genomes, lie in the same or different genes. The test is carried out by arranging for the two genomes to be present in the same cell (or organism, the test applies throught the living world). If the wild-type phenotype results, the mutations were in different genes; if the mutant phenotype results, then the mutations were in the same gene. This is usually found to be a difficult concept; there is not space here to go into it in more detail, but it is worth thinking about!

Problem 26	**Specialized transduction with phage λ**

Question

Strain X100, described in Problem 21, is *galK⁻*; it is also lysogenic for the temperate phage λ, which integrates into the *E. coli* chromosome at a unique site close to the *galETK* operon. In the *prophage* state, λ, like many temperate phages, can be *induced*, i.e. appropriate treatments (such as, in this case, low doses of ultraviolet radiation) cause it to leave the dormant prophage state and enter the *lytic cycle*, with eventual cell lysis and release of new phage particles.

A culture of strain X100 was induced in this way and the resulting λ phage particles collected. These were then allowed to infect a further strain X110, which was not lysogenic for λ and whose only genetic marker was an uncharacterized *gal* mutation (i.e. it was not known in which gene of the operon the mutation lay). The infected cells were plated on Min + gal (i.e. minimal medium with galactose as sole carbon source, on which X110 forms only microcolonies); a small number of normal-sized colonies grew, corresponding to about 1 per 10^5 λ particles.

Some of these were then induced in the same way. The resulting lysates were collected and allowed to infect two different strains: (i) X110 and (ii) X111, which had a complete deletion of the *gal* operon. The mixtures of cells and phage particles were finally plated on Min + gal. Normal-sized colonies resulted in all three cases, the approximate average numbers of colonies being: (i) 1 per 5 λ particles and (ii) 1 per 20 λ particles.

1. What was the genetic constitution of the normal-sized colonies on Min + gal from the first X110 mixture?
2. What genetic events could explain the numbers of normal-sized colonies on Min + gal from the second set of mixtures?

Answer

1. Evidently 1 in 10^5 λ particles from the induced X100 carry genetic information that, when expressed in X110, produce a Gal$^+$ phenotype. This arises from the position of the λ prophage close to the *gal* operon (Fig. 26.1); the *integrated* state of λ is achieved by a single crossover between specific attachment sites on the phage and bacterial chromosome, much as shown for integration of F in Fig. 24.1. Induction results in *excision*, normally the exact reverse of the integration event (Fig. 26.2(a)) but occasionally *aberrant* (Fig. 26.2(b)) so that the λ genome carries, in place of part of its own DNA, a chromosomal DNA segment that can include *gal*. The resulting genome is termed $\lambda dgal$; the *d* stands for 'defective', but some genomes originating in this way are non-defective, in which case they are termed *p* for 'plaque-forming' (this will not be discussed further here). If this genome had been formed in a *gal$^+$* host, the *gal* DNA now attached to λ would have the wild-type sequence; we could then write $\lambda dgal^+$. In this case, however, the host was *gal$^-$*, so we write $\lambda dgal^-$. The $\lambda dgal$ genome will eventually be *packaged* into a phage head, which will in yturn be assembled into a phage particle. The $\lambda dgal$-carrying phage particle may then adsorb to a further *E. coli* host cell and inject its genome in the usual way; this type of process, whereby bacterial DNA is carried

Figure 26.1.
Generation of the specialized transducing genome $\lambda dgal$. The lysogenic strain X100, showing integrated λ prophage next to the *galETK* operon.

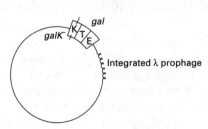

Figure 26.2. Excision of λ can proceed either normally (a), or via aberrant excision via an illegitimate recombination (b), exactly as with F in an Hfr (see Problem 16 and Figure 24.4).

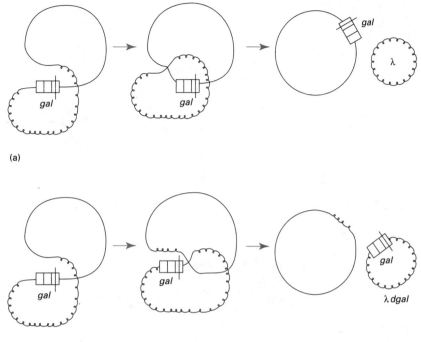

(a)

(b)

from one strain to another within a phage head, is termed *transduction*. Once inside its new host, the λ*dgal* genome will usually integrate by homologous recombination between the chromosomal segment and its counterpart on the recipient chromosome: note that the chromosomal segment found in the λ*dgal* is present twice, i.e. the transductant is a *partial diploid* for this segment. What will be the *gal* genotype of the transductant? The possible results are shown in Fig. 26.3.

Even if the donor segment and recipient chromosome both carry *gal⁻* mutations, this situation can (as long as the mutations do not involve the same base pairs) give rise to Gal⁺ colonies. This can happen by either *complementation* or recombination, as follows. Integration can occur by recombination either (i) between the *gal⁻* mutations (Fig. 26.3(a)i, (b)i) or (ii) between one or other of these and the end of the duplicated segment (Fig. 26.3(a)ii, (b)ii). If the *gal⁻* mutations are in different genes, but *not* if they are in the same gene, then (ii) will give rise to complementation (see Problem 25); (i), on the other hand, will result in a wild-type *gal⁺* form of the gene anyway.

This form of transduction is termed *specialized transduction*; specialized because only donor markers closely linked to an inserted prophage can be transferred (compare generalized transduction and transformation, described in Problem 27). The Gal⁺ colonies are referred to as (specialized) *transductants*.

2. Specialized transductants are usually lysogenic simultaneously for λ*dgal* and for wild-type λ, because the original lysate consisted of mostly normal λ with a small proportion of λ*dgal*. Induction of these specialized transductants therefore gives a mixture of particles with the normal λ genome and of ones with λ*dgal* genomes. There are usually rather more of the former because they apparently replicate

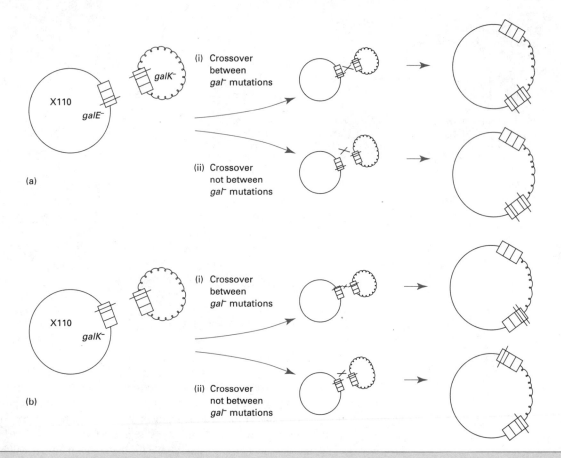

Figure 26.3. The alternatives when the λ*dgal*⁻ from X100 infects X110. (a) X110's *gal*⁻ mutation is in *galE* or *galT*. In both (i) and (ii) there are two copies of the *gal* operon flanking the λ prophage. In (i), one of these is wild-type and the other carries both mutations (i.e. in *galE* and *galK*). In (ii), the two copies each carry one of the original mutations. As these are different genes, complementation occurs. (b) X110's *gal*⁻ mutation is in *galK*, as with X100. The same comments apply to (i) and (ii) as given in part (a), but this time as the original mutations in (ii) are in the same gene, complementation cannot occur.

better. The frequency of specialized transductants is therefore vastly higher than the first time round. Why is the number lower for X111? Complementation cannot happen with X111: there is no *gal* DNA in the chromosome to complement the *gal*⁻ in the λ*dgal*⁻, as it has all been deleted. These particular transductants must therefore originate from recombination events that generated λ*dgal*⁺, either in the first integration (Fig. 26.3(a)i, (b)i) or following induction.

Problem 27

Transformation three-point cross

Question

The bacterium *Bacillus subtilis* can, when in an appropriate physiological state (termed *competent*), take up DNA from the medium, a process known as *transformation*. If this

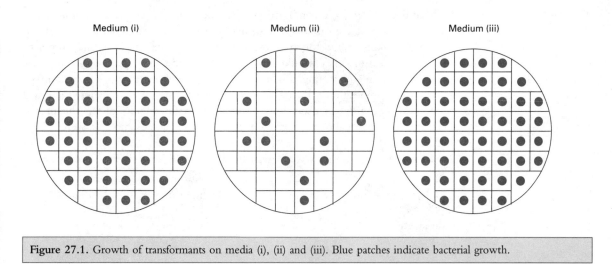

Figure 27.1. Growth of transformants on media (i), (ii) and (iii). Blue patches indicate bacterial growth.

DNA consists of sheared (and therefore linear) fragments of chromosomal DNA from *B. subtilis* or a closely related species, it can recombine with the host chromosome to generate recombinants, as occurred following conjugation in Problems 21, 22 and 23. In one such experiment, the *donor* strain BS200 (strain from which the DNA fragments were obtained) was prototrophic and carried a mutation *tmp* that made it resistant to the antibiotic trimethoprim (tmp). The *recipient* strain BS201 (strain that took up the DNA) carried mutations *ilvD* and *thyB* that made it auxotrophic for isoleucine + valine (ilv) and thymine (thy) (note that wild-type *B. subtilis* is exactly like *E. coli* in its ability to grow on minimal salts plus glucose medium) and had the wild-type sensitivity to antibiotics. The mixture of competent BS201 cells and fragments of BS200 DNA was plated on a complete medium containing tmp that permitted auxotrophic cells to form colonies.

The resulting transformants were *patched out* in a grid pattern on the same complete medium + tmp to give *master plates*, which were then *replica plated* on to the following media: (i) Min + ilv, (ii) Min + thy, and (iii) complete + tmp. The replica plates from one such master plate are depicted in Fig. 27.1; each grid position shows either growth or non-growth.

1. What genotypes can grow on plates (i), (ii) and (iii)? What is the purpose of medium (iii)?
2. By comparing each grid position in turn on plates (i), (ii) and (iii), tabulate the genotypes of all the transformants on this master plate.
3. What are the possible arrangements of the three markers? On the basis of each, explain how each transformant genotype arises.
4. From the fact that one of the genotypes is much rarer than the other three, deduce which of the arrangements in (3) is the correct one.
5. What are the genetic distances between the markers?

Answer

1. Medium (i): *thy*⁺, any combination of the other markers; medium (ii): *ilv*⁺, any combination of the other markers; medium (iii): *tmp*ʳ, any combination of the other markers. Since the transformants were selected on the same complete medium + tmp, they would be expected to grow on medium (iii) anyway; however, in replica plating, it is held necessary to replicate finally on to medium of the same composition as the master plate, to ensure that all places where there was growth on the master plate show growth on the final replica plate. This proves that if test plates fail to show growth, this is not just because no cells were left on the replicating material!

2. This is best done as shown below, where + denotes growth and − no growth.

	Medium (i)	Medium (ii)	Medium (iii)
Grid position 1	+	−	+
2	+	+	+
etc.			

3. There are three possible arrangements, with each marker in turn being the middle one. In each arrangement, the markers define four regions of DNA, and the four possible transformant genotypes *thy*⁺ *ilv*⁺, *thy*⁺ *ilv*⁻, *thy*⁻ *ilv*⁺ and *thy*⁻ *ilv*⁻ (remember that the transformants are selected for *tmp*ʳ) result from various crossover patterns in these regions (Fig. 27.2).

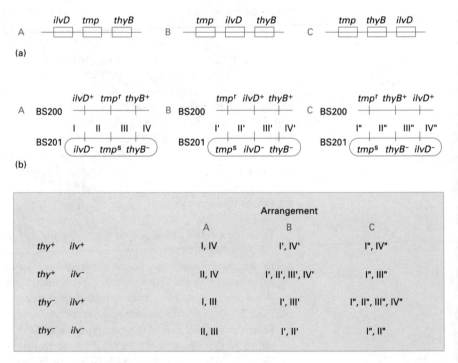

Figure 27.2. Crossover patterns required to give the various recombinant classes for the three possible gene orders shown in (a). (b) Shows the arrangements of markers and chromosomal regions in which crossovers can occur. The four transformant genotypes for the *thyBilvD* markers would originate from the crossover patterns in (c), according to the three gene arrangements.

4. If *tmp* is the middle gene (Fig. 27.2, gene order A), all four transformant classes can be accounted for by two crossovers (note that any recombination that occurs must regenerate the 'circular' bacterial chromosome: this means that there must be an even number of crossovers between the chromosome and the incoming fragment). However, for the two other arrangements of markers (Fig. 27.2, gene orders B, C), one transformant class requires four crossovers, which would be expected to be appreciably less frequent than the others. The two arrangements make different predictions as to which transformant class will be the least frequent; if *thyB* is the middle gene, then $thy^- \, ilv^+$ is the least frequent class, whereas if *ilvD* is in the middle, then $thy^+ \, ilv^-$ is the least frequent. In fact, $thy^- \, ilv^+$ is the least frequent class, which proves that the order of genes is *tmp–thyB–ilvD*. (In a more advanced treatment, it could be allowed that there are certain provisos to this reasoning.)

5. The genetic distance between two markers is measured by the recombination frequency between them. So the genetic distance between *tmp* and *thyB* is estimated here as the frequency of crossovers between tmp^r (which was selected) and thy^+, namely $(1+5)/52 = 12\%$ (1 is the number of $thy^- \, ilv^+$ transformants, 5 the number of $thy^+ \, ilv^-$ transformants; in both classes a crossover has occurred between *tmp* and *thyB*, in region II of Fig. 27.2b). Similarly, the genetic distance between *tmp* and *ilvD* is the frequency of crossovers between tmp^r and ilv^+. This might be thought to be $(33+5)/52 = 73\%$; this however overlooks the fact that in the $thy^- \, ilv^+$ transformant, *two* crossovers have occurred in this interval, in regions II and III of Fig. 27.2b. Hence the true genetic distance is $(1 \times 2 + 33 + 5)/52 = 77\%$. The genetic distance between *thyB* and *ilvD* is estimated as the proportion of transformants in which a crossover has taken place between these two genes; this has occurred in $thy^- \, ilv^+$ and $thy^+ \, ilv^-$ transformants, so giving $(1+33)/52 = 65\%$. Note that the distances are *additive*, i.e. $12\% + 65\% = 77\%$.

Although this question refers to DNA transfer by transformation, exactly the same reasoning applies to *generalized transduction*, in which random fragments of donor DNA are transferred to a recipient bacterial cell within a virus particle. Similar considerations apply also to specialized transduction, although the partial diploid nature of the transductants is a complicating factor; generalized transductants are haploids, not partial diploids like specialized transductants.

Finally, this problem has shown how marker order can be determined simply, i.e. without any arithmetic, by the criterion that the correct order correctly predicts the least frequent recombinant class, as arising through a multiple crossover pattern. The same criterion can be used in mapping by conjugation (Problems 21–23), in phage crosses (Problem 28), and in crosses involving eukaryotes (Problems 16–18).

Problem 28

Phage three-point cross

Question

Phage T4 is a virulent phage of *E. coli*. As with other phages, most genes encode indispensable functions, so that loss of function is lethal. Two kinds of mutation that

Type	Appearance of plaque	Growth on X400 at 30 °C	Growth on X400 Sup at 42 °C	Number of plaques
1	rI^+	No	Yes	28
2	rI	Yes	No	28
3	rI^+	No	No	15
4	rI	Yes	Yes	15
5	rI^+	Yes	No	5
6	rI	No	Yes	5
7	rI^+	Yes	Yes	2
8	rI	No	No	2

Table 28.1. Number of plaques grown on X400 or X400 Sup media.

can be used are (i) temperature-sensitive mutations, and (ii) suppressible mutations. *Temperature-sensitive* mutations lead to the gene's protein product being thermolabile; the phage then develops normally when the bacterial host is grown at low temperatures, but fails to develop when it is grown at higher temperatures. *Suppressible* mutations lead to the gene's protein product normally having one or more amino acid differences from the wild type that render it non-functional in normal bacterial hosts. Certain hosts may however carry *suppressor* mutations in transfer RNA genes; the mutant transfer RNAs then *mistranslate* the mutant codons in the messenger RNA, such that amino acids are inserted which permit the protein to function.

Two genetically marked phage strains were crossed, as follows. Strain 300 carried a temperature-sensitive mutation in gene *A* (written *Ats*), so that it developed normally in *E. coli* hosts at 30 °C but not at 42 °C. It also carried a mutation in gene *rI*. The *rI* product is dispensable and such mutants show merely an altered *plaque* morphology: whereas rI^+ plaques are small and fuzzy at the edge, *rI* mutant plaques are larger and sharp to the edge. Strain 301 carried a suppressible mutation in gene *E* (the mutation was of the so-called 'amber' type, ·so it was written *Eam*); this mutant could not develop in the wild-type *E. coli* host X400 but could in strain X400 Sup, which carried a suppressor of 'amber' mutations.

Two high-titre stocks of strains 300 and 301 were added together and separately to a culture of *E. coli* X400, which was then incubated at 42 °C. The separate phage stocks failed to show lysis of the culture even after several hours, while the mixture lysed the culture within 1 hour, as does a wild-type phage stock. The progeny phages released from cells infected with the mixture were allowed to form individual plaques on a lawn of X400 Sup at 30 °C. Phages from 100 of the plaques were then tested on (i) lawns of X400 with incubation at 30 °C and (ii) lawns of X400 Sup with incubation at 42 °C. The appearance of each plaque, i.e. whether this corresponded to an rI^+ or an *rI* phenotype, was also noted. The numbers of the eight possible types of progeny phage are shown in Table 28.1.

1. Explain why the mixture of mutant phage stocks lysed the host cells rapidly, while each mutant stock on its own failed to do so at all.
2. What are the genotypes of the eight classes of progeny phage?
3. Which marker is the middle one?
4. What are the genetic distances between the markers?

Answer

1. The *rI* mutation does not affect the ability of the phage to lyse infected cells in the normal way, so why does co-infection by one phage strain carrying a temperature-sensitive mutation and another carrying a suppressible mutation, both apparently in genes encoding indispensable functions, give productive lytic infection at the higher growth temperature of a bacterial host strain not carrying a suppressor? (Note that 'temperature-sensitive mutation' is actually an incorrect term; it should be 'temperature-sensitivity mutation', since it is not the mutation that is temperature sensitive but the gene product – most people would regard this as overly pedantic.) The answer is that this is another example of complementation. Phage 300, which carries *Ats*, also carries the wild-type allele E^+ of the *E* gene, which gives rise to functional wild-type E protein. Phage 301, which carries *Eam*, also carries the wild-type allele A^+ of the *A* gene, which gives rise to functional wild-type A protein. Hence when X400 is infected by a mixture of 300 and 301, wild-type E and A proteins are present and functional (the presence of non-functional Eam and Ats mutant forms notwithstanding) and the lytic cycle proceeds just as if X400 had been infected with wild-type phage.

2. type 1: rI^+ *Eam* A^+
   type 2: *rI* E^+ *Ats*
   type 3: rI^+ *Eam Ats*
   type 4: *rI* E^+ A^+
   type 5: rI^+ E^+ *Ats*
   type 6: *rI Eam* A^+
   type 7: rI^+ E^+ A^+
   type 8: *rI Eam Ats*

3. The same type of reasoning applies as in Problem 27. There are three possible arrangements of the genes, each one in turn being in the middle. In this case, however, recombination occurs between two full-length phage genomes (not between a 'circular' bacterial chromosome and a small linear fragment, as in transformation or generalized transduction with incoming chromosomal DNA). Consequently, any number of crossovers is permitted, whether even or odd. Of the eight genotypes listed in (2), two are parental (types 1 and 2), four involve a single crossover, and two, the least frequent, involve two crossovers. One should work out what crossovers give rise to each genotype, and which genotype would be least frequent, for each of the three gene orders. It turns out that only the assumption that *E* is in the middle predicts types 7 and 8 to be the least frequent; since this is in fact the case, this assumption must be correct, i.e. the gene order is *rI–E–A*.

4. Again, the reasoning of Problem 27 applies. The genetic distance between *rI* and *E* is calculated as follows: a single crossover has occurred between these two markers in types 5, 6, 7 and 8, so that the genetic distance is given by $(5 + 5 + 2 + 2)/100 = 14\%$. For the genetic distance between *E* and *A*, a single crossover has occurred between these two markers in types 3, 4, 7 and 8, so that the genetic distance is given by $(15 + 15 + 2 + 2)/100 = 34\%$. Finally, for the genetic distance between *rI* and *E*, a single crossover has occurred between these two markers in types 3, 4, 5 and 6, and two crossovers in types 7 and 8, so that the genetic distance is given by $(15 + 15 + 5 + 5 + 2 \times [2 + 2])/100 = 48\%$. Note that as in Problem 27 the distances are additive, i.e. $14\% + 34\% = 48\%$.

THE PHYSICAL MAPPING OF DNA

Introduction

Three techniques are central to the physical mapping of DNA.

The first is the use of electrophoresis to separate DNA fragments according to size. In a charge field, small DNA fragments are more mobile than large fragments, and so the size of uncharacterized fragments can be estimated by comparing their mobility with standard fragments of known size.

The second technique is restriction endonuclease digestion. Endonucleases are enzymes that cut DNA internally. Restriction endonucleases have very specific sequence requirements for cleavage and some cut DNA very precisely wherever their specific target sequence occurs. Different restriction endonucleases may have different target sequences.

The third technique is DNA hybridization. This specific recognition of the two complementary strands of DNA can be exploited to detect whether sequences identical, or very similar, to a known probe occur in a separated fragment of DNA. The probe DNA may, for example, be radioactively labelled and added to samples of DNA fragments generated by restriction endonuclease digestion. The procedure used normally involves transferring samples of DNA fragments to a special membrane to which the DNA binds (the technique of Southern blotting) and then adding probe DNA, which will hybridize to fragments with matching sequences.

Using these three techniques it is possible to locate where particular restriction endonuclease target sequences occur in a DNA molecule and so to map them. The location of genes for which probes are available can then be determined.

The location of restriction endonuclease target sequences proceeds as follows. Suppose that a particular target sequence occurs only once in a circular plasmid molecule. Digestion with the relevant enzyme (enzyme A) will cut the DNA at that sequence, generating a linear molecule from the circle (this result is also detectable by electrophoresis since linear DNA molecules are more mobile than circles). Now suppose that a different restriction endonuclease has two target sequences in the plasmid. After digestion with this enzyme (enzyme B), two linear fragments will be produced, the size of which can be determined by electrophoresis. For a plasmid of 10 kilobase pairs (kbp), we might for example find fragments of 4 kbp and 6 kbp. It is now possible to relate the site cut by enzyme A to those cut by enzyme B. The plasmid molecule is digested with a mixture of the enzymes A + B. If we find that fragments of 1, 4 and 5 kbp are generated following double digestion, we must conclude that the target site for enzyme A lies within the 6-kbp fragment generated by enzyme B. We can piece this together and draw a map of the plasmid as shown in Fig. 29.1.

We can now probe samples of these fragments with the DNA sequence of, for example, a gene coding for resistance to an antibiotic, which is 1 kbp in length. If we detect hybridization between the probe and both the 1 and 5 kbp fragments, we may conclude that the gene straddles the target sequence for enzyme A. Using these

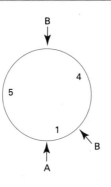

Figure 29.1. A map of the plasmid following double digestion.

procedures it is possible to build up complex physical maps. The series of problems that follow start simply but illustrate how more complex maps can be built up.

Problems 29–33 all involve the same plasmid (circular DNA molecule), which is 8000 base pairs (= 8 kbp) in length. The plasmid contains two genes coding for resistance to antibiotics, one for ampicillin resistance, the other for hygromycin resistance. Problem 34 involves a plasmid being transformed into a recipient, and the different types of integration event analysed by restriction analysis and Southern blotting.

Problem 29	### Simple restriction mapping of a plasmid with two enzymes

Question

When the plasmid is digested by the restriction endonuclease, *Eco*RI, two fragments, sized 2 and 6 kbp are produced. When digested by a second restriction endonuclease, *Hind*III, two fragments are also produced but these are 1.5 and 6.5 kbp in size. Digestion with a mixture of the two enzymes yield four fragments, sized 0.5, 1, 1.5 and 5 kbp. Draw a map of the plasmid, representing the physical relationship of the enzyme target sites.

Answer

We may first conclude that since each enzyme gives rise to two fragments upon digestion of the plasmid DNA, there must be two target sites in the plasmid for each enzyme. Furthermore, since the double digestion yields four distinct fragments, the target sites for the two enzymes must be well separated.

To determine how the target sites of the two enzymes relate to one another, we must reconcile the results of the single digestions with those from the double digestion. Starting with the fragments generated with *Eco*RI alone, and relating these to the double digest, we must identify the fragments from the double digest that have come from the two *Eco*RI fragments. This can only be done in one way. The 2-kbp fragment from *Eco*RI digestion must give rise to the 0.5 and 1.5 fragments on double digestion and so one target site for *Hind*III must lie within the 2-kbp *Eco*RI fragment. Similarly the 6-kbp *Eco*RI fragment must generate the 1 and 5 kbp fragment upon double digestion and so the other *Hind*III target site must lie within the 6-kbp *Eco*RI fragment. We can piece this information together as shown in Fig. 29.2.

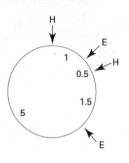

Figure 29.2. A map of the plasmid showing restriction sites and fragment sizes.

This map is consistent with the fragments generated both by the single digests and the double digest (check this for yourself!).

Problem 30	**Further restriction mapping of a plasmid – incorporating a third enzyme**

Question

The plasmid is digested with a third restriction endonuclease, *Pst*I, generating fragments of size, 2.5 and 5.5 kbp. Digestion with a mixture of *Eco*RI and *Pst*I gives fragments sized 1.5, 2 and 2.5 kbp, and with a mixture of *Hind*III and *Pst*I gives fragments sized 1, 1.5, 2.5 and 3 kbp. Digestion with all three enzymes together generates fragments sized 0.5, 1, 1.5 and 2.5 kbp.

Draw a new map of the plasmid including the *Pst*I target sites as well as those for *Eco*RI and *Hind*III.

Answer

Note first that once again the single digest with *Pst*I generates two fragments, so it too has two target sites within the plasmid. The *Eco*RI + *Pst*I double digest generates only three fragments. Since each of these enzymes has two target sites, it is to be expected that four fragments would be generated by the double digestion. There are two explanations for finding only three fragments.

1. One of the target sites for one of the two enzymes may lie very close to one of the target sites for the other enzyme. Upon digestion a very small fragment would be generated and such fragments are difficult to detect following electrophoretic separation.
2. The expected four fragments may be generated but two are, by chance, so nearly similar in size that they have not been separated by electrophoresis.

It is easy to distinguish between these explanations by considering the sum of the sizes from the double digest, which is $1.5 + 2 + 2.5 = 6$ kbp. Since the plasmid is 8 kbp, two fragments sized 2 kbp must have been generated by the double digest to account for the apparent shortfall; thus we need to reconcile the production of fragments sized

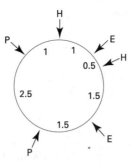

Figure 30.1. A restriction map of the plasmid, showing restriction sites for the three enzymes and fragment sizes between restriction sites.

1.5, 2, 2 and 2.5 kbp with the map we have already established in the answer to Problem 29. We must conclude that there is no target site for *Pst*I within the 2-kbp fragment generated by *Eco*RI digestion, and that both *Pst*I target sites therefore lie within the part of the plasmid between the two *Eco*RI sites which are separated by 6 kbp of DNA. From the information given so far we cannot determine the order of the *Pst*I sites within this 6-kbp sequence.

Double digestion with *Hin*dIII and *Pst*I gives the expected four fragments with a combined length equal to the whole plasmid; thus the target sites for the two enzymes are well separated and no pair of fragments are similar in size. To reconcile the products of double digestion with those from digestion with each enzyme singly, we must see how the fragments from the double digestion can be grouped to generate those from the single digests. There is only one way that this is possible: the 6.5-kbp fragment given by *Hin*dIII digestion must give rise to the 1, 2.5 and 3 kbp fragments from double digestion, i.e. both *Pst*I target sites must lie within this 6.5-kbp sequence. However, once again we are unable to decide from the data so far considered the order of the *Pst*I sites within the *Hin*dIII-generated fragment.

Digestion with all three enzymes generates only four fragments but we know there should be six since we have already deduced that all the target sites are well separated. We also observe that the sum of the fragment sizes from triple digestion is 5.5 not 8 kbp as expected, i.e. that 2.5 kbp of sequence is unaccounted for. The missing two fragments and the 2.5 kbp of DNA must be accounted for by the generation of fragments of nearly similar size, which are not separated by electrophoresis. The two extra fragments must be 1 and 1.5 kbp in size. The triple enzyme digestion therefore generates 1×0.5, 2×1, 2×1.5 and 1×2.5 kbp fragments.

We can now try to fit all the new information relating to the *Pst*I sites to the map already established in the answer to Problem 29. We have concluded that both *Pst*I sites occur within the 6-kbp sequence generated by *Eco*RI, dividing it to give 1.5, 2 and 2.5 kbp fragments, and within the 6.5-kbp fragment generated by *Hin*dIII, dividing it to give 1, 2.5 and 3 kbp fragments. Thus both *Pst*I sites must lie within the 5-kbp sequence generated by *Eco*RI + *Hin*dIII double digestion. Furthermore, one *Pst*I site must lie 1 kbp away from the *Hin*dIII site shown at the top of the plasmid map (Fig. 29.1), since in this position both the 1-kbp fragment from *Hin*dIII + *Pst*I double digestion and 2-kbp fragment from *Eco*RI + *Pst*I double digestion are accounted for. The other *Pst*I site must lie 1.5 kbp away from the *Eco*RI site towards the bottom right of the map in order to generate both the 1.5-kbp fragment from the *Eco*RI +

*Pst*I double digestion and the 3-kbp fragment from the *Hind*III + *Pst*I double digestion. We can check these deductions. The two *Pst*I target sites should be 2.5 kbp apart, which they are (the *Hind*III, *Eco*RI fragment in which they both lie is sized 5 kbp, one site is 1 kbp from one end, the other is 1.5 kbp from the other end). The complete map is therefore as shown in Fig. 30.1.

You should now carefully check this map with the results. The map predicts fragments as follows:

Single digests:	*Eco*RI	2 and 6 kbp
	*Hind*III	1.5 and 6.5 kbp
	*Pst*I	2.5 and 5.5 kbp
Double digests:	*Eco*RI + *Hind*III	0.5, 1, 1.5 and 5 kbp
	*Eco*RI + *Pst*I	1.5, 2 (×2) and 2.5 kbp
	*Hind*III + *Pst*I	1, 1.5, 2.5 and 3 kbp

The fragments from the triple digest have the sizes shown in Fig. 30.1, and so the map is consistent with the data.

Problem 31

Yet more simple restriction mapping – locating a unique restriction site on a restriction map

Question

Digestion of the plasmid with the endonuclease *Bam*HI gives rise to a single linear DNA molecule, 8 kbp in size, indicating that there is only one target site for this enzyme in the plasmid. Digestion with *Bam*HI + *Eco*RI yields fragments 1.5, 2 and 4.5 kbp. Predict the possible outcomes of digestion with a mixture of *Bam*HI and either *Hind*III or *Pst*I, and explain how you would use these to locate the *Bam*HI target site with respect to the other sites.

Answer

Since 1.5 and 3.5 kbp fragments result from the digestion of the plasmid with *Bam*HI + *Eco*RI, the target site for *Bam*HI must lie within the 6-kbp sequence generated by *Eco*RI digestion. The *Bam*HI site could therefore be at either of two positions arrowed 1 and 2 in Fig. 31.1 (note that one of the positions must be very near to a *Pst*I target site).

Figure 31.1. The location of the unique *Bam*HI site (1) on the restriction map of the plasmid.

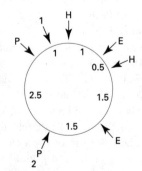

We can predict the outcome of digestion of the plasmid with a mixture of either *BamHI* + *HindIII* or with *BamHI* + *Pst*1. The predictions will be different for the two possible locations of the *BamHI* target site..

BamHI site at:	1	2
Digestion with:		
BamHI + *HindIII*	0.5, 1.5, 6	1.5, 3, 3.5
BamHI + *PstI*	0.5, 2.5, 5	2.5, 5.5

These different predictions therefore provide us with a way of deciding which of the two sites is the correct one.

Problem 32

Identification of a specific fragment in a restriction map by probing

Question

Double digestion with *BamHI* + *HindIII* gave fragments sized 0.5, 1.5 and 6 kbp.

A 1-kbp probe containing the coding sequence for the ampicillin-resistance gene is hybridized with the fragments generated by digestion of the plasmid with each of the restriction endonucleases singly and in pairs. The results are summarized in Table 32.1.

Answer

The digestion with *BamHI* + *HindIII* tells us that position 1 in the map given in the answer to Problem 31 represents the correct location of the *BamHI* target sequence.

The results of hybridization between the probe for the ampicillin-resistance gene and the fragments generates more information than is needed to locate the gene, but it is always useful to have extra information for confirmation. Of the digestions with single restriction endonucleases, the results with the fragments generated by *Pst*I are the most informative. Since both fragments hybridize with the probe, one of the two *Pst*I target sequences must be located inside the ampicillin-resistance gene. Digestion

Table 32.1. Fragment hybridization with ampicillin-resistance probe.

Restriction endonuclease	Hybridize (kbp)	Do not hybridize (kbp)
*Eco*RI	6	2
*Hind*III	6.5	1.5
*Pst*I	2.5, 5.5	–
*Bam*HI	8	–
*Eco*RI + *Hind*III	5	0.5, 1, 1.5
*Eco*RI + *Pst*I	2, 2.5	1.5
*Eco*RI + *Bam*HI	1.5, 4.5	2
*Hind*III + *Pst*I	1, 2.5	1.5, 3
*Hind*III + *Bam*HI	0.5, 6	1.5
*Pst*I + *Bam*HI	0.5, 2.5, 5	–

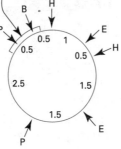

Figure 32.1. The ampicillin-resistance gene located on the plasmid restriction map.

approximate position of ampicillin-resistance gene

with a mixture of *Pst*I and *Bam*HI also generates fragments all of which hybridize with the probe. We can therefore conclude that the *Bam*HI target sequence must also be within the gene and so the ampicillin gene must lie towards the top left of our map. (Had the gene contained the other *Pst*I site, then the 0.5-kbp fragment generated by digestion with *Pst*I + *Bam*HI would not have hybridized.) We can therefore redraw our map of the plasmid to contain this information (Fig. 32.1).

You should check through the results of hybridization with the fragments generated by other digestions to ensure that the suggested location of the ampicillin gene is consistent with these other results.

Problem 33

Further probing for specific restriction fragments

Question

Finally, to locate the position of the gene coding for hygromycin resistance, a 1.8-kbp probe containing the coding sequence of this gene was hybridized with the fragments generated by single and double digestions with the four restriction endonucleases, to give the results shown in Table 33.1. Complete the plasmid map by indicating the location of the coding region of the hygromycin-resistance gene.

Table 33.1. Fragment hybridization with hygromycin-resistance probe.

Restriction endonuclease	Hybridize (kbp)	Do not hybridize (kbp)
*Eco*RI	2, 6	–
*Hind*III	1.5, 6.5	–
*Pst*I	5.5	2.5
*Bam*HI	8	–
*Eco*RI + *Hind*III	0.5, 5, 1.5	1
*Eco*RI + *Pst*I	1.5, 2	2.5
*Eco*RI + *Bam*HI	2, 4.5	1.5
*Hind*III + *Pst*I	1.5, 3	1, 2.5
*Hind*III + *Bam*HI	1.5, 6	0.5
*Pst*I + *Bam*HI	5	0.5, 2.5

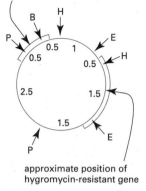

approximate position of
ampicillin-resistance gene

approximate position of
hygromycin-resistant gene

Answer

The result of hybridization with the fragments generated by the *Eco*RI restriction endonuclease shows that, since both fragments hybridize with the probe, then one or other of the two target sites for this enzyme must be contained in the hygromycin-resistance gene's coding region. The similar finding with the fragments generated by *Hin*dIII digestion likewise indicate that one or other of the *Hin*dIII target sequences must be within the hygromycin-resistance gene. The gene must therefore be located in the region of these four target sequences (upper right on our map), but can we be more precise? Look at the map we have already built up (Fig. 32.1) and consider which double digest will generate information that will enable us to locate the gene more precisely. Note that the *Eco*RI + *Pst*I double digest is not helpful because we do not know whether the hybridization with a 2-kbp fragment involves only the fragment of this size generated by *Eco*RI alone or in addition the same-sized fragment released by double digestion. The *Eco*RI + *Bam*HI double digest is more informative. Since the 4.5-kbp fragment does hybridize with the probe but the 1.5-kbp fragment does not, we may conclude that the hygromycin-resistance gene's coding sequence spans the *Eco*RI and *Hin*dIII target sequences, which are 1.5 kbp apart.

The final map is therefore as shown in Fig. 33.1.

Problem 34

Analysis of types of transformant integration events

Question

We are transforming a uridine-requiring *pyr-4* strain of *Neurospora crassa* with the cloned *pyr-4* gene in plasmid pAR*pyr*. The donor gene is wild type and the recipient gene is mutant, but the point mutant in the recipient strain does not alter the restriction sites for the two enzymes used in this work, *Eco*RI and *Bam*HI.

In pAR*pyr*, the entire *pyr-4* is located in a single *Bam*HI fragment 2 kbp in size, inserted in the single *Bam*HI site of the 8-kbp plasmid vector pAR1. There is an internal *Eco*RI site that cuts the *Bam*HI fragment into 1.5 and 0.5 kbp. When pAR*pyr* is cut with *Eco*RI, it produces two fragments 6.0 and 4.0 kbp in length. The

		Band size (kbp)									
DNA	Probe	1	2	3	4	5	6	7	8	9	10
Genomic	pAR1										
	pARpyr	+	+								
	pyr-4	+	+								
pAR1	pAR1								+		
	pARpyr								+		
	pyr-4										
pARpyr	pAR1				+		+				
	pARpyr				+		+				
	pyr-4				+		+				
trans 1	pAR1			+		+	+				
	pARpyr	+	+	+		+	+				
	pyr-4	+	+			+	+				
trans 2	pAR1				+			+	+		
	pARpyr	+	+		+			+	+		
	pyr-4	+	+		+			+			
trans 3	pAR1				+		+				
	pARpyr	+	+		+		+				
	pyr-4	+	+		+		+				
trans 4	pAR1				+					+	+
	pARpyr	+	+		+					+	+
	pyr-4	+	+		+						+
trans 5	pAR1										
	pARpyr	+	+								
	pyr-4	+	+								

Table 34.1. DNA fragments identified by hybridization with the probe specified (pAR1, pARpyr and the isolated pyr-4 BamHI fragment).

EcoRI/BamHI double digest of pARpyr gives four fragments, 4.5, 3.5, 1.5 and 0.5 kbp.

The resident pyr-4 gene and its chromosomal environs were analysed, and probing with a cloned pyr-4 gene identified a single fragment of 2.0 kbp on a Southern blot of genomic DNA cut with BamHI, two fragments of 1.0 and 2.0 kbp on an EcoRI digest and two fragments of 1.5 and 0.5 kbp on a BamHI/EcoRI double digest.

The Neurospora recipient strain is transformed and pyrimidine-independent transformants isolated. Five of the transformants are isolated for further study. As controls, pAR1, pARpyr and untransformed Neurospora genomic DNA are used. Table 34.1 gives the results for these and the four transformants after cutting the DNA completely with EcoRI, running it out on an agarose gel by electrophoresis, Southern blotting it onto nitrocellulose membrane and probing with the radioactively labelled probes specified.

1. Identify the homologous and heterologous integrants.
2. For the homologous types, identify the conversion and tandem types.

3. For each transformant, draw a restriction map of the cloned *pyr-4* gene and its position relative to other hybridizing fragments.

4. For each heterologous integrant, state the size(s) of any *de novo* boundary fragment(s) generated by the integration event.

Answer

In the filamentous ascomycete fungi it is possible to achieve genetic transformation, but the fate of transforming DNA is always to integrate into a chromosomal location.

Integration into a chromosomal site is possible by both homologous and heterologous events. In the former, either 'gene conversion' results in replacement of the resident (defective) gene sequence by the donor (functional) one without any integration of vector sequence, or a single crossover between the resident and donor copies of the gene results in a tandem duplication of the gene, with the vector sequence between the two copies.

From the data given, the maps of the vector plasmid pAR1 and plasmid pAR*pyr* (both circular, but drawn as linear molecules cut at *Eco*RI site) and of the resident *pyr-4* locus in the linear chromosome are shown in Fig. 34.1.

Five transformants are studied. As controls, pAR1, pAR*pyr* and untransformed *Neurospora* genomic DNA are used. The patterns of hybridization to vector, plasmid and genomic DNA are shown in Fig. 34.2.

In a conversion or double crossover homologous transformant, the restriction map of the chromosomal region containing the *pyr-4* gene will not change. Hence the results of probing with pAR1, pAR*pyr* and *pyr-4* will be the same as for the untransformed recipient strain, the 1-kbp and 2-kbp *Eco*RI bands (Fig. 34.2a). Transformant 5 shows this result.

Single crossover homologous insertion and ectopic insertion will both give transformant DNA that will hybridize with a pAR1 probe. The former will also give

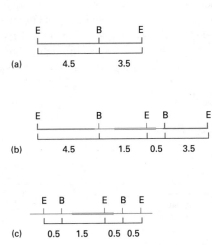

Figure 34.1. (a) Restriction map of vector plasmid pAR1. (b) Restriction map of plasmid pAR*pyr*. (c) The resident *pyr-4* gene and its chromosomal environs.

(a)

E   B   E

4.5   3.5

(b)

E   B   E   B   E

4.5   1.5   0.5   3.5

(c)

E   B   E   B   E

0.5   1.5   0.5 0.5

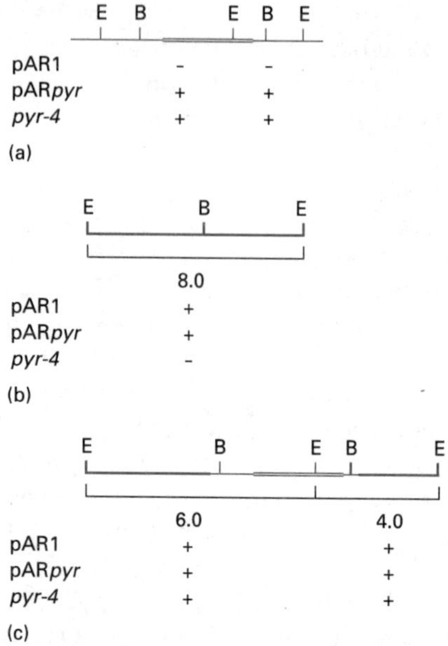

Figure 34.2. The interpretation of data in Table 34.1 showing which fragments hybridize to the three probes (+) and which do not (−). (a) Probing of genomic DNA cut with *Eco*RI. (b) Probing of vector (pAR1) DNA with *Eco*RI. (c) Probing of plasmid (pAR*pyr*) DNA cut with *Eco*RI.

the same two *Eco*RI fragments as the pAR*pyr* vector (4.0 and 6.0 kbp) plus the same two fragments as the recipient strain (1.0 and 2.0 kbp). The isolated *pyr-4* probe will also highlight all four bands, as shown in Fig. 34.3. This is characteristic of transformant 3, a single crossover homologous insertion.

However, ectopic transformants will show the resident 1 and 2 kbp fragments with pAR*pyr* and *pyr-4* probes, plus either the 4 or 6 kbp plasmid fragment (but not both) depending on the site in the vector involved in the insertion, and two new junction fragments of sizes depending on the position in the plasmid of the exchange and the flanking *Eco*RI sites of the host chromosome. Thus transformants 1, 2 and 4 are ectopic. In Fig. 34.4, for each ectopic transformant, the resident copy is shown to the left and the ectopically integrated copy to the right, with their respective restriction maps and patterns of hybridization to the probes.

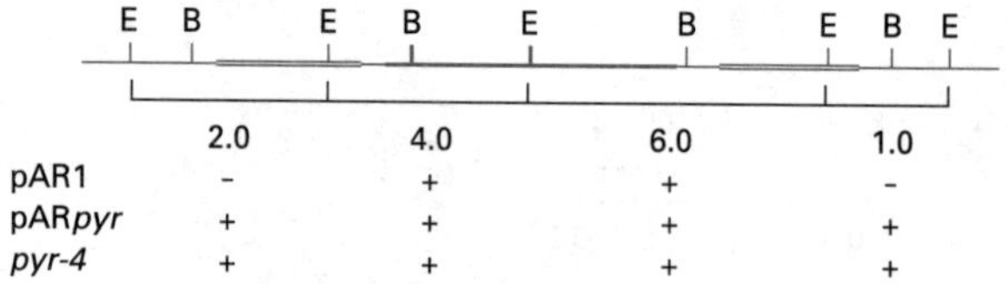

Figure 34.3. Restriction map and hybridization pattern of transformant 3.

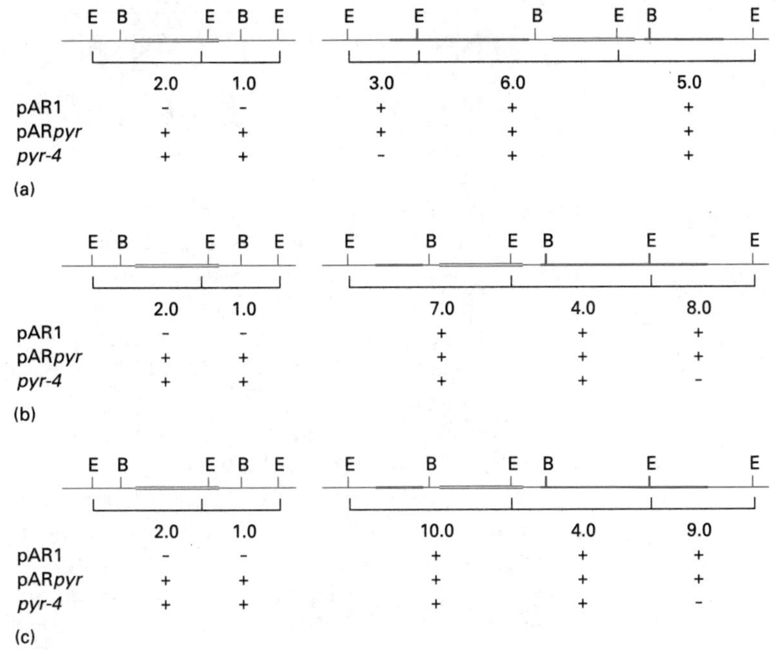

Figure 34.4.
Restriction map and hybridization patterns of (a) transformant 1; (b) transformant 2; (c) transformant 4.

	2.0	1.0		3.0	6.0		5.0
pAR1	–	–		+	+		+
pAR*pyr*	+	+		+	+		+
pyr-4	+	+		–	+		+

(a)

	2.0	1.0		7.0	4.0	8.0
pAR1	–	–		+	+	+
pAR*pyr*	+	+		+	+	+
pyr-4	+	+		+	+	–

(b)

	2.0	1.0		10.0	4.0	9.0
pAR1	–	–		+	+	+
pAR*pyr*	+	+		+	+	+
pyr-4	+	+		+	+	–

(c)

USING DELETION MUTANTS TO MAKE A GENETIC MAP

Introduction

Deletion mutants lack part of the normal DNA sequence. The region deleted may vary greatly in size from a single base pair to millions of base pairs. Mutants having deletions of more than a few base pairs are very stable, since it is impossible to regain the lost genetic information by further mutation. Deletion mutants are therefore a very reliable source of material for the construction of genetic maps. Their principal utility, however, lies in the fact that they provide a qualitative rather than a quantitative method of mapping the genetic material. Suppose we have two different deletion mutants each of which has part of the same gene missing. We cross the two mutants together and see whether they can give rise to recombinant progeny that have inherited the whole gene intact (and is therefore functional). If we are able to recover such progeny, we may conclude that the regions of the gene deleted in the two mutants did not overlap. However, if no such recombinant progeny can be recovered, we may conclude that, on the contrary, the regions deleted in the two mutants must overlap, at least partially. These contrasting conclusions can be drawn without recourse to counting the number of recombinant progeny. If we obtain recombinants that have regained a functional gene, then the deletions do not overlap; if we do not obtain recombinants within an adequate sample of progeny, they do overlap.

How do we use these conclusions to construct a map of a gene? Let us begin with a map of a gene divided arbitrarily into segments a–l as follows:

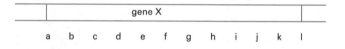

Suppose that deletion mutant X1 is missing segments a, b, c and d and deletion mutant X2 is missing segments i, j, k and l. When these two mutants are crossed they can recombine anywhere in the region e–f–g–h to give rise to progeny that have the entire gene sequence since mutant X1 can provide the DNA sequence missing in mutant X2 and vice versa. However, if mutant X1 is crossed to a different mutant, X3, which lacks segments d, e and f, no recombinants with the complete gene can be produced since both mutants lack segment d. Note that these crosses do not provide information of the size of the deleted regions, only of their relationship to one another. Thus two deletions, one of which (mutant X4) is missing the whole gene except for segment l and another (mutant X5) missing just segment a, will behave the same qualitatively when crossed to mutant X6, which lacks just segment l.

There are various conventions for representing deletion mutants. A common one is to highlight the region of the gene missing in the mutant. We would represent mutant X1 using this convention as follows:

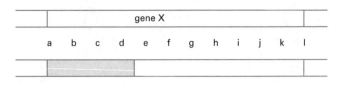

We could try and represent more accurately that mutant X1 is missing part of the DNA sequence of the gene, for example as follows:

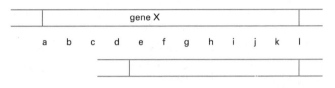

However this convention makes it difficult to represent the way in which different deletion mutants relate to one another. A convenient compromise convention, where a gap is left corresponding to the missing sequence, is as follows:

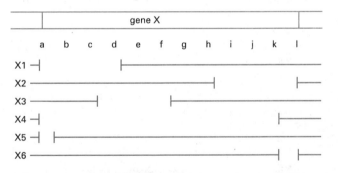

Using this convention we can now draw a map, representing all the deletions that have been used as examples so far:

When constructing deletion maps, it is not always possible to resolve, at least initially, ambiguities relating to the order of deletions. Let us consider a new set of deletion mutants of gene Y, mutants Y1–5. We can represent the results of crosses between these mutants in a table, where + indicates that the two mutants can give rise to progeny having a complete gene and − where they cannot.

	Y2	Y3	Y4	Y5
Y1	−	−	−	−
Y2		−	+	+
Y3			−	+
Y4				+

From this information we cannot decide between two possible maps:

Map A

Map B

Note the reversed order of deletions 2 and 4 in the two maps. The choice between these two maps may be resolved if further data involving other deletion mutants are available. For example, if another deletion, Y6, gives recombinants having the complete gene only with mutant Y4, then Map B is the correct one:

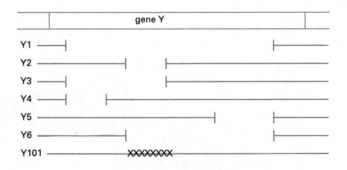

Deletion maps of this sort, once established can be used to rapidly locate point mutants (mutants having a single base pair mutated). Suppose the set of deletion mutants for gene Y illustrated above are each crossed to a new point mutant, Y101, and it is found that recombinant progeny having a functional gene Y are recovered only in the crosses between Y101 and mutants Y4 and Y5, then we may deduce that the point mutant is changed somewhere in the region highlighted in the map above.

The series of problems that follow will give you practice in constructing genetic maps using deletion mutants.

Problem 35

Elementary intra-genic deletion mapping

Question

Five deletion mutants (*lac*Δ 1–5) of the *Escherichia coli lacZ* gene are crossed by transduction and progeny are screened for recombinants able to synthesize β-galactosidase (the product of the *lacZ* gene). The results are summarized in Table 35.1 (+, recombinants recovered able to make β-galactosidase; −, no such recombinants recovered). Draw a map showing the relationship between the regions of the *lacZ* gene deleted in these mutants.

Table 35.1. Results of inter-crossing the *lacZ* deletions.

	Δ2	Δ3	Δ4	Δ5
Δ1	−	+	+	+
Δ2		+	−	−
Δ3			−	+
Δ4				−

Answer

Deletion mutant 4 fails to give recombinants having an intact *lacZ* gene with either deletion mutant 2 or 3, so the deleted sequence in Δ4 must overlap that deleted in both Δ2 and Δ3. Since Δ2 and Δ3 do give rise to recombinants with *lacZ* function, the sequences deleted in these two mutants must not overlap. We can represent this diagrammatically, using the convention we have discussed in the introduction to this chapter (Fig. 35.1). Notice that any of the maps in Fig. 35.1b–e are equally valid interpretations of the data given for Δ2, Δ3 and Δ4 alone.

Deletion mapping can only establish the mutual overlaps (and lack of them) between deletion mutants and all the maps shown are similar in this respect. Understanding this is crucial to the development of the deletion map as more data become available, since this will usually necessitate modification of the map but only so as to alter the pattern of overlaps so far established. Thus Δ1 does not overlap Δ4 but does overlap Δ2, and so the map in Fig. 35.1c must be along the right lines. Taking into account all the data, the map in Fig. 35.2 can be drawn.

However it must be emphasized again that this is only a diagrammatic representation of the relationships between the deletions and cannot, with the data available, be related to the gene itself. The whole map may require compressing to relate to some small portion of the gene, and may have to be reversed with respect to the genes flanking the *lacZ* gene.

Figure 35.1. (a) A possible deletion map of *lacZ*. (b)–(e) Alternative *lacZ* deletion maps.

Figure 35.2. A refined deletion map of *lacZ*.

Problem 36 — Deletion mapping using both deletions and point mutants

Question

rII mutants of bacteriophage T4 are unable to grow on *E. coli* strain K12λ. Such mutants are affected in one of two genes, *rIIA* and *rIIB*. Four deletion mutants affecting the *A* gene, Δ1–4, have been obtained. Deletion mutants Δ1, Δ2 and Δ3 extend past the right-hand border of the *A* gene into the adjacent *B* gene. The deletion in mutant Δ4 is confined to the *A* gene. Mutants 101–106 are the result of point mutation in the *A* gene. The four deletion mutants were crossed to one another and also to the six point mutants, and the progeny viruses obtained were tested for growth on *E. coli* K12λ. The results are summarized in Table 36.1 (+, *some* progeny obtained that are able to grow on *E. coli* K12λ; −, no such progeny obtained).

Draw a map of the *rIIA* gene indicating the extent of the deletions and the relative locations of the region affected by the point mutations.

Table 36.1. Results of inter-crossing deletions Δ1–4 with deletions Δ1–4 and point mutants 101–106.

	Δ2	Δ3	Δ4	101	102	103	104	105	106
Δ1	−	−	−	+	−	+	−	+	−
Δ2		−	+	+	+	+	−	+	+
Δ3			−	−	−	−	−	+	−
Δ4				−	+	+	+	+	−

Answer

We need to note first that the extent of the deletions Δ1, Δ2 and Δ3 into the *rIIB* gene is unknown. The data given only enable deductions to be made relating to the *rIIA* gene. Since point mutant 105 can recombine with all of the deletion mutants to give progeny having a complete *rIIA* gene, this mutant must be affected in the *rIIA* gene towards the end furthest from the *rIIB* gene. Mutants 101 and 103 must be affected somewhere in the middle of the gene, mutants 102 and 106 more towards the *rIIB* end of the *rIIA* gene and mutant 106 must be affected nearer to the *rIIB* gene (it fails to give recombinants with *rIIA* function with any of the deletions extending into the *rIIB* gene). At this stage we can draw a map incorporating these conclusions, as shown in Fig. 36.1.

The results with the internal deletion of the *rIIA* gene, mutant Δ4, allow us to develop this map further. Since Δ4 gives recombinants with Δ2, it must have a right-hand border to the left of Δ2's left-hand border. Since it gives functional *rIIA* recombinants with point mutant 103 but not 101, the left border of the deletion must lie between the affected sites in these two mutants and 103 must map to the left of

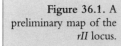

Figure 36.1. A preliminary map of the *rII* locus.

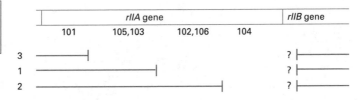

Figure 36.2. A deletion map of the *rII* locus.

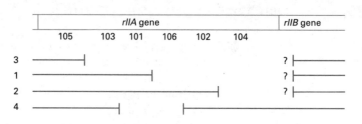

101. Furthermore, since Δ4 recombines with 102 but not 106, the right border of the deletion must lie between the sites of these two point mutants, with 102 mapping to the right of 106. These deductions can be incorporated into the map shown in Fig. 36.2.

You should now go back to the data and check the results of the crosses against the map that has been deduced. Remember that where the deletions of two mutants overlap no recombination to form an intact gene is possible, and where a point mutant is affected in the DNA sequence that is missing in a deletion mutant, it is again impossible for recombination to generate an intact gene.

Problem 37

Deletion analysis in a eukaryote

Question

In the fungus, *Aspergillus nidulans*, the *nii* and *nia* genes are adjacent on chromosome 8. Deletion mutants affecting both genes are readily identifiable since loss of function of both genes occurs as a result of a single mutation. A series of such mutants, D1–10, are crossed with a number of *nia* point mutants (P1–10) and the ability to recover recombinant progeny with functioning *nia* genes is scored to give the data in Table 37.1.

Construct a map of the *nia* gene based upon these data showing the extent of the deletions in each of the mutants D1–10 and the relative location of the point mutants P1–10.

Table 37.1. Results of inter-crosses between point mutants P1–10 and deletions D1–10.

	P1	P2	P3	P4	P5	P6	P7	P8	P9	P10
D1	−	−	−	+	+	+	−	−	−	−
D2	−	+	−	+	+	+	−	−	+	−
D3	−	+	+	+	+	+	+	−	+	+
D4	−	−	−	−	−	−	−	−	−	−
D5	−	−	−	+	+	−	−	−	−	−
D6	+	+	+	+	+	+	+	−	+	+
D7	−	−	−	−	+	−	−	−	−	−
D8	−	−	−	+	+	+	−	−	+	−
D9	−	+	−	+	+	+	−	−	+	+
D10	−	+	+	+	+	+	−	−	+	+

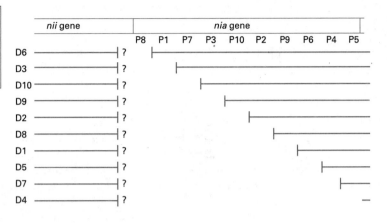

Figure 37.1. Map of the overlapping deletions located with reference to the point mutants.

Answer

Note first that all the deletions affect two adjacent genes, and therefore all must extend to delete the border between the two genes and hence the same end of the *nia* gene. The data in this problem do not relate to the extent of the deletions into the *nii* gene. Since point mutant P8 fails to give recombinants with any of the deletion mutants, we can conclude that this must lie nearest to this (left-hand) border. Since P5 gives recombinants with all the deletions except D4, this point mutation must lie furthest to the right of our map, and deletion D4 must extend at least as far as the site of P5 and possibly have the entire *nia* gene deleted. We can extend this logic to take account of the rest of the data and construct a map of the *nia* gene (Fig. 37.1).

In this map, the same convention for representing deletion mutants as has been used elsewhere in this chapter has been used, i.e. the line represents the DNA sequence that is present in the deletion mutant strain. The left-hand border of each deletion cannot be determined and so is shown with a question mark against its location in the *nii* gene. While the deleted sequence must extend into the *nii* gene, it may affect anything from a few base pairs to the whole *nii* gene.

You should now check that the results given in the problem are consistent with the above map.

MUTATION ANALYSIS

5

Introduction

The very basis of genetics is the process of mutation and the mutants caused by that process. Without strain differences resulting from mutation, generating different alleles at particular loci that are expressed as different phenotypes, there would be no genetics. Of course mutation can occur spontaneously, but geneticists have been using artificially induced mutants for decades, initially induced by radiation but subsequently by chemical mutagens with different spectra of effects on DNA. Any genetics textbook will provide details of physical and chemical mutagens and their effects on DNA.

The simplest mutants are base pair substitutions, where a particular base is replaced by another base. Base pair substitutions may be transitions (purine for purine and pyrimidine for pyrimidine) or transversions (purine for pyrimidine and vice versa). In coding DNA in an open reading frame (that part of the DNA translated), a base pair substitution may cause a substitution of a single amino acid for another, but because of redundancy in the genetic code may cause no change at the level of the translated polypeptide. At the extreme, with mutation from a coding codon to a stop codon, the result may be the more serious premature termination of translation, resulting in a truncated peptide with a normal amino-terminus but a truncated carboxy-terminus.

A base pair substitution other than a chain termination one (which results in an amino acid substitution) may have no effect on the gene product, may alter the properties of the gene product (e.g. lower the thermostability) or may inactivate the gene product. All except the first would have a detectable phenotype, at least under some conditions. All except the first may well alter the conformation of the gene product, so it may have different antigenic properties.

Other point mutants result in the insertion or deletion of one or a few base pairs. If this occurs in a coding region and the insertion or deletion is not three base pairs or a multiple thereof, the consequence will be a frameshift, in which the coding frame will be shifted and the message misread from that point on until a misread codon results in chain termination. Unless at the extreme carboxy-terminus of the gene product, frameshifts will result in total inactivation of the gene product.

Inactive mutants may be further mutagenized and secondary mutants with some restoration of activity may be selected. These might be of two types, allelic (in the same gene as the original mutant) or non-allelic. Allelic revertants of base pair substitutions that cause amino acid substitutions may alter the same codon to restore the original coding specificity or replace with yet another amino acid that may be more acceptable. However, allelic revertants may make a compensatory base pair substitution at another codon in the gene: allelic revertants of base pair substitution chain termination mutants will alter the stop codon to an acceptable sense codon; allelic revertants of frameshift mutants will insert or delete one or a few base pairs to restore the reading frame. Frameshift revertants may be near or at the site of the original frameshift. If the former, the region between may differ from the original sequence as long as it contains no chain termination codons or inserts no unacceptable amino acid residues.

Non-allelic 'reversion' is more properly called 'suppression'. There are non-allelic nonsense suppressors that occur at tRNA genes, giving mutant tRNA molecules which can translate a stop codon. Other types of non-allelic suppression involve a variety of mechanisms, including altering regulatory mechanisms, producing re-routing of metabolites into other pathways, etc.

In Problem 38, the consequences of frameshift and allelic suppression are examined. This will involve the use of a diagram of the genetic code, shown below.

UUU	phe	UCU	ser	UAU	tyr	UGU	cys
UUC	phe	UCC	ser	UAC	tyr	UGC	cys
UUA	leu	UCA	ser	UAA	stop	UGA	stop
UUG	leu	UCG	ser	UAG	stop	UGG	trp
CUU	leu	CCU	pro	CAU	his	CGU	arg
CUC	leu	CCC	pro	CAC	his	CGC	arg
CUA	leu	CCA	pro	CAA	gln	CGA	arg
CUG	leu	CCG	pro	CAG	gln	CGG	arg
AUU	ile	ACU	thr	AAU	asn	AGU	ser
AUC	ile	ACC	thr	AAC	asn	AGC	ser
AUA	ile	ACA	thr	AAA	lys	AGA	arg
AUG	met	ACG	thr	AAG	lys	AGG	arg
GUU	val	GCU	ala	GAU	asp	GGU	gly
GUC	val	GCC	ala	GAC	asp	GGC	gly
GUA	val	GCA	ala	GAA	glu	GGA	gly
GUG	val	GCG	ala	GAG	glu	GGG	gly

Note that in the DNA, U is replaced by T

Problem 39 involves mutation and reversion, but integrates these data with data from recombination and complementation. The recombination is based on the material covered in Chapter 1, involving determination of independent assortment and linkage, and the degree of any linkage. It also involves complementation, which has not been covered elsewhere in this book.

Generally speaking, the vast majority of mutant alleles are found to be recessive to their wild equivalents, most because the mutants have lost activity of the gene and/or its product. If two haploid strains are mutant in two different genes, and are brought together in a diploid or heterokaryon (so that for each gene, the original and mutant alleles of each are present in the same cell), the wild-type (if functional and dominant) allele of each gene will be expressed. This will result in a phenotype no different from wild type. In this situation, we have complementation between the two differentially mutant strains. The ability of two mutant strains to complement is taken to mean that the mutants are in different genes (cistrons, coding units) and hence are not allelic. There are rare combinations of alleles of the same gene in which some interaction of subunits in a multimeric protein gene product permits a low level of activity and complementation, but that is beyond the scope of this book.

Problem 38	**The use of suppressed frameshift mutants in determining DNA sequence from protein sequence**

Question

The fungus *Neurospora crassa* is able to grow on cellulose as its sole carbon source by producing extracellular cellulase enzymes that hydrolyse polymeric carbohydrate to glucose. Mutants deficient in this enzyme activity are not lethal to the organism as long as another carbon source is available. Growth on medium containing cellulose and another carbon source results in a low but detectable level of the cellulase enzyme (by its activity on cellulose).

The major cellulase enzyme has been purified, sequenced and antibodies against it produced. The antibody preparation reacts, as expected, with the cellulase antigen in spent medium.

An acridine-induced mutation has resulted in a derivative strain with no detectable cellulase activity and thus unable to grow on cellulose as sole carbon source. The specific antibody against the cellulase does not react with any antigen in the medium in which this strain has been grown.

In a subsequent mutation experiment, a derivative of this mutant is selected in which cellulase activity has been restored, thus 'reverting' the initial mutation and restoring the antigenic protein. The cellulase of this revertant was then purified and the amino acid sequence determined.

The original and revertant proteins were compared and found to differ in one short sequence near the active site indicated in bold below.

Wild type
leu thr trp gln lys cys thr arg trp cys pro thr leu asn thr thr met

Revertant
leu thr **cys arg ser ala pro gly gly ala arg pro cys** asn thr thr met

1. What was the precise nature of the original mutation event and what was the nature of the reversion mutation?
2. As far as possible, what is the DNA sequence for the wild-type gene in this region?

Answer

The initial mutation eliminated the activity of the enzyme and either eliminated the protein or changed it significantly so that it was no longer similar to the wild type. This could be a major deletion, a chain termination point mutation or a missense mutation that significantly altered the configuration of the gene product.

Because the mutation is 'revertible', we can deduce that it was not a gross mutation such as a deletion of any significant part of the gene. Only point mutants normally revert.

Because a series of 10–11 amino acid residues differ between the original wild type and the revertant, we can deduce that this is neither a reverting missense mutant (in which either one or two amino acid residues would differ) nor a reverted or suppressed 'nonsense' (chain termination) mutant, in which a single amino acid residue might differ. It is in fact a frameshift mutant, reverting by a compensatory frameshift near it in the DNA sequence but not at exactly the same residue.

If we put in the DNA sequence (including alternatives) of the sense strand for the wild type and revertant, we can deduce the mechanism:

Wild-type amino acid sequence

leu	thr	trp	gln	lys	cys	thr	arg	trp	cys	pro	thr	leu	asn	thr	thr	met
TTA	**ACT**	TGG	CAA	AAA	TGT	ACT	AGA	**TGG**	TGT	CCT	ACT	TTA	**AAT**	**ACT**	**ACT**	**ATG**
TTG	ACC		CAG	AAG	TGC	ACC	AGG		TGC	CCC	ACC	TTG	AAC	ACC	ACC	
CTT	ACA					ACA	CGT			CCA	ACA	CTT		ACA	ACA	
CTC	ACG					ACG	CGC			CCG	ACG	CTC		ACG	ACG	
CTA							CGA					CTA				
CTG							CGG					CTG				

Revertant amino acid sequence

leu	thr	cys	arg	ser	ala	pro	gly	gly	ala	arg	pro	cys	asn	thr	thr	met
TTA	**ACT**	TGT	AGA	TCT	GCT	CCT	**GGT**	**GGT**	GCT	AGA	CCT	TGT	**AAT**	**ACT**	**ACT**	**ATG**
TTG	ACC	TGC	AGG	TCC	GCC	CCC	GGC	GGC	GCC	AGG	CCC	TGC	AAC	ACC	ACC	
CTT	ACA		CGT	TCA	GCA	CCA	GGA	GGA	GCA	CGT	CCA			ACA	ACA	
CTC	ACG		CGC	TCG	GCG	CCG	GGG	GGG	GCG	CGC	CCG			ACG	ACG	
CTA			CGA	AGT						CGA						
CTG			CGC	AGC						CGC						

A comparison of the two sequences above, starting from the left end (N-terminal for the protein, 5′ for the sense strand of the DNA), shows them to be identical for the first eight residues but that the G in position 8 or 9 is lacking in the reverted mutant. Furthermore, working back from the other end, a compensatory base (T or C) has been inserted at position 39 in the revertant, beyond which the sequences are identical. Highlighted in bold in the two sequences are the identical possible codons at both ends, and between positions 9 and 39 the possible codons in the revertant and how they could have been derived from the possible codons in this part of the sequence in the wild type.

It is not possible to say from the data given, for questions (1) and (2), which was the original frameshift and which the compensatory (suppressor) frameshift. However, whichever way round the two mutation events occurred, the result is a region of 10-amino acids translated out of frame. The nature and location of this difference is insufficient to inactivate the enzyme, although the original frameshift (with just a single base insertion or deletion) would have been inactive due to a totally non-functional sequence downstream from the site of the mutation.

Before direct DNA sequencing, suppressed frameshifts were the only way of reducing the codon ambiguity deduced from protein sequencing. Using this method, reducing the codon ambiguity by eliminating those theoretical codons in the wild-type sequence that are not compatible with those required to code for the amino acid sequence in the suppressed frameshift, considerable refinement of the sequence is possible. The only ambiguity remaining in the region between the compensatory frameshifts is that for the arginine codon, as shown below:

Wild-type amino acid sequence

leu	thr	trp	gln	lys	cys	thr	arg	trp	cys	pro	thr	leu	asn	thr	thr	met
TTA	ACT	TGG	CAG	AAG	TGC	ACC	AGG	TGG	TGC	CCG	ACC	TTG	AAT	ACT	ACT	ATG
TTG	ACC						CGG					CTG	AAC	ACC	ACC	
CTT	ACA													ACA	ACA	
CTC	ACG													ACG	ACG	
CTA																
CTG																

Problem 39 — Determining the nature of mutation events

Question

In *Saccharomyces cerevisiae*, 10 different adenine-requiring mutants (M1–M10) were isolated from a mutation experiment. The mutant selection experiment was carried out at 37 °C throughout. Adenine auxotrophs are readily detected as, apart from the inability to grow without adenine in the medium, they accumulate red pigment in the cells.

The 10 mutant strains and a wild-type control were sub-cultured onto duplicate plates of minimal medium and the two sets were incubated at 34 and 37 °C respectively. After 2 days, it was observed that the wild type had grown at both temperatures, mutant M3 had grown at 34 °C and the other mutants had failed to grow at either temperature.

The 10 mutants were then tested for complementation. All pairwise combinations were inoculated onto minimal medium and the plates incubated at 37 °C for 2 days. Using + for growth and − for no growth, the results of this complementation test showed:

	M1	M2	M3	M4	M5	M6	M7	M8	M9
M10	+	+	+	+	−	+	+	+	−
M9	−	+	+	−	−	+	+	+	
M8	+	+	−	+	+	+	+		
M7	+	−	+	+	+	−			
M6	+	−	+	+	+				
M5	+	+	+	+					
M4	−	+	+						
M3	+	+							
M2	+								

To investigate possible linkage, all pairwise combinations were crossed on supplemented medium and the resulting spores plated and germinated on supplemented and minimal medium. Samples of 100 ascospores from each cross were germinated and grown on supplemented medium and the numbers of those that grew when sub-cultured onto minimal medium are shown below:

	M1	M2	M3	M4	M5	M6	M7	M8	M9
M10	0	12	23	0	0	8	11	28	0
M9	0	9	24	0	0	11	9	24	
M8	27	22	0	23	27	26	22		
M7	9	0	22	12	9	0			
M6	12	0	26	8	9				
M5	0	11	27	0					
M4	0	8	23						
M3	22	26							
M2	12								

Reversion tests were then carried out on the mutants, using the mutagens nitrosoguanidine (which causes mainly base pair substitutions) and ICR-170 (an acridine compound causing mainly frameshifts). The results of these were:

	Nitrosoguanidine	ICR-170
M1	+	−
M2	+	−
M3	+	−
M4	−	+
M5	+	−
M6	−	+
M7	+	−
M8	+	−
M9	−	−
M10	−	+

Finally, the 10 mutants were crossed with a phenotypically wild-type strain (FB3) which, on crossing with wild type, was known to give 25% uridine auxotrophs. The uridine auxotroph that segregated was revertible with nitrosoguanidine but not with ICR-170. The results of these crosses were:

	$ade^-\ uri^-$	$ade^-\ uri^+$	$ade^+\ uri^-$	$ade^+\ uri^+$
FB3×M1	12	14	12	62
FB3×M2	11	39	14	36
FB3×M3	14	35	11	40
FB3×M4	12	38	14	36
FB3×M5	13	37	10	40
FB3×M6	14	39	12	35
FB3×M7	13	38	12	37
FB3×M8	13	14	10	63
FB3×M9	11	35	15	39
FB3×M10	15	37	12	36

Deduce as much as you can from the above data about the nature of the mutational changes in the mutants, the number of genes, their functions and their linkage if any.

Answer

First of all, any conditional mutant, such as the temperature-sensitive M3, is almost certainly a base pair substitution resulting in a near-normal protein with an unstable secondary structure. The other mutants have an absolute requirement for adenine and could be of various types. Further data are necessary to determine their nature.

The complementation test results allow us to sort the different alleles into several groups. M3 and M8 will not complement each other but will complement all other mutants, so are both in the same complementation unit (cistron). Mutants M2, M6 and M7 do not complement each other, but do complement all other mutants; they therefore represent a second complementation unit. Mutants M1 and M4 define a third complementation unit, not complementing each other but complementing M2, M3, M5, M6, M7 and M10. Mutants M5 and M10 define a fourth complementation unit, complementing M1, M2, M3, M4, M6 and M7. Mutant M9 overlaps the third and fourth complementation units, behaving like a deletion lacking at least part of both these cistrons. In summary, we have defined four cistrons: cistron A1 (M1, M4), cistron A2 (M5, M10), cistron B (M2, M7 and M8) and cistron C (M3 and M8). Mutant M9 affects cistrons A1 and A2.

The results of the inter-crosses show if two mutants are separable by recombination at a level that should detect recombination between all except very closely linked genes. If unlinked, the expected result would be 25% wild type, 25% each of the two single mutant types and 25% of the double mutant, i.e. 25% prototrophic and 75% auxotrophic. If within the same gene or very closely linked, few if any wild-type recombinants would be expected. If the inter-crossed mutants were in two linked genes, wild-type recombinants at a level significantly less than 25% might be expected. In the actual results, no wild-type recombinants are found from crosses between alleles within cistron A1, cistron A2, cistron B or cistron C. No wild-type recombinants result from crosses between alleles in cistron A1 with alleles within cistron A2, nor from mutant M9 with either cistron A1 or cistron A2. Therefore this defines three recombinational loci: locus A (mutants M1, M4, M5, M9 and M10), locus B (equivalent to cistron B) and locus C (equivalent to cistron C). Crosses between locus A and locus C give 25% wild-type recombinants, indicative of independent assortment. Crosses between locus B and locus C give 25% wild-type recombinants, indicative of independent assortment. Crosses between locus A and locus B give 10% wild-type recombinants, indicative of linkage at 20 map units.

The reversion analysis suggests that mutants M4, M6 and M10 are frameshifts, as they revert with ICR-170 but not nitrosoguanidine. Mutants M1, M2, M3, M5, M7 and M8 are base pair substitutions, as they revert with nitrosoguanidine but not ICR-170. The failure of M9 to revert at all is consistent with the earlier suggestion that it is a deletion.

The crosses to the phenotypically wild-type strain that segregated the uridine allele (which by its reversion behaviour was a base pair substitution) shows the Mendelian expectation of a 1 : 1 segregation of the adenine allele to be perturbed in the crosses with mutants M1 and M8, but in all other crosses a 1 : 1 segregation results. The results of the M1 and M8 crosses suggest that in 25% of the progeny the adenine requirement is being suppressed. The strain with a 'wild-type phenotype' appears to be carrying a uridine mutant suppressed by a non-allelic, unlinked, suppressor gene. This suppressor is capable of suppressing two alleles (M1 and M8) at the adenine loci A and C as well as the uridine mutant allele. As we know that the uridine mutant is a base pair substitution, the suppressor is a nonsense suppressor and is misreading and translating potential chain termination mutants at the uridine, adenineA and adenineC loci. With both mutant M1 at locus A and mutant M8 at locus C, 25% of the progeny of the cross to the suppressor require adenine for growth, so neither locus A nor locus C is linked to the suppressor locus. Further, the results of the crosses show that neither locus A nor locus C is linked to the uridine locus.

Mutants M2, M5 and M7 revert in a way characteristic of base pair substitutions, but are not suppressible by the nonsense suppressor strain. A given nonsense suppressor strain only suppresses one of the three chain termination types (UAA, UAG or UGA). These three mutants may therefore be either nonsense (of a different type to M8) or missense base pair substitutions.

In summary, the results define three adenine loci, A and B linked at 20 map units and C unlinked. Locus A is complex, with two cistrons. None of the adenine loci is linked to the suppressor locus, nor to the uridine locus, and the suppressor and uridine loci are not linked to each other.

The 10 adenine auxotrophs can be classified for locus and cistron, type of change (base pair substitution, frameshift or major deletion) and nature of base pair substitution (missense or nonsense):

Allele	Locus	Type
M1	A1	Chain termination (nonsense) base pair substitution
M2	B	Base pair substitution
M3	C	Missense conditional base pair substitution
M4	A1	Frameshift
M5	A2	Base pair substitution
M6	B	Frameshift
M7	B	Base pair substitution
M8	C	Chain termination (nonsense) base pair substitution
M9	A1, A2	Deletion
M10	A2	Frameshift

HUMAN GENETICS

6

Introduction

It is not always possible to carry out experimental crosses to establish the pattern of interitance shown by a character. This is obviously true of human characters but also applies to some animals such as racehorses. In these cases, pedigrees are often available and these can be used to establish the basis of the inheritance of the character, provided that this is simple. Pedigrees use symbols to represent individuals, circles for females, squares for males. The symbol is shaded to indicate that an individual has the character in question, and left unshaded if the individual is unaffected. Matings are shown by joining symbols with a horizontal line, and one or more vertical lines from this horizontal line join parents to offspring. Examples of these symbols are given below.

○ unaffected female
● affected female
■ unaffected male
□ affected male

mating between unaffected female and affected male

○——■

We will consider here the results of only three simple patterns of inheritance. In each, a condition is caused as a result of mutation in a single gene. The mutant allele may be recessive or dominant to the normal allele. If it is dominant, then we would expect every individual who carried the mutant allele to be affected. We can therefore deduce that affected children should always have at least one of their parents affected, and the corollary that two unaffected parents should not have children who are affected. Where the mutant allele is recessive, pedigrees will show two contrasting patterns, depending on whether the affected gene is on the X chromosome or one of the other chromosomes (autosomes).

For autosomal recessive conditions, both copies of a gene must be mutant for an individual to be affected. An affected individual may however have two unaffected parents, since each parent may have one mutant allele and one normal allele, i.e. be heterozygous, and so not show the condition. Such unaffected heterozygous individuals are often referred to as carriers.

Where a condition results from a recessive allele of a gene carried on the X chromosome a sex-linked pattern of inheritance will be observed. Since human males have only one X chromosome, they will show the condition if this carries a mutant allele. To be affected, a female must have recessive mutant alleles on both of her X chromosomes. Since male offspring inherit their X chromosome from their mother, an affected mother will have sons who all show the condition. If the mother is a carrier, then there will be an equal chance that a son will be affected or unaffected.

Problem 40	**Simple human pedigree analysis**

The pedigrees shown in Fig. 40.1 are conditions whose inheritance is caused by mutation in a single gene. One of the conditions is caused by the possession of an autosomal dominant allele. The other two pedigrees involve conditions caused by the possession of recessive alleles, one autosomal, the other sex-linked. Which pedigree relates to each of these causes? Deduce the genotypes of the individuals in the pedigrees as fully as possible.

Pedigree C in Fig. 40.1 is consistent with the condition being caused by the possession of an autosomal dominant allele. Note that all affected individuals have an affected parent and that the mating between two unaffected individuals gives only unaffected offspring. If we call the dominant mutant allele A and the corresponding normal (recessive) allele a, we can allocate genotypes to individuals as shown in Fig. 40.2.

Note that:
1. All unaffected individuals must be homozygous for the recessive allele (a/a).
2. Affected individuals that have unaffected offspring must be heterozygotes (i.e. have one recessive allele to pass on to their unaffected offspring). Thus both III.10 and III.11 must be A/a since they have a/a offspring.
3. The genotypes of individuals IV.1 and IV.3 may be either A/A or A/a.

Pedigree B in Fig. 40.1 is consistent with the condition being caused by homozygosity of a recessive allele. If we designate the normal (dominant) allele A and the recessive allele, which when homozygous leads to the condition, a, we can allocate genotypes as shown in Fig. 40.3.

Note that:
1. If homozygosity for a recessive allele leads to a condition, we can only be sure of the genotypes of affected individuals (a/a).
2. We can, however, deduce that if unaffected parents have an affected offspring, then both parents must be heterozygotes (A/a).
3. The mating III.3 and III.4 is between first cousins. It is not certain that they each received their recessive allele from their related parents (II.2 and II.7) but if the condition is rare in the population at large, then this is much more likely than either the male, II.1, or the female, II.8, being heterozygous.

Pedigree A in Fig. 40.1 is consistent with the condition being caused by a recessive allele carried on the X chromosome, i.e. a sex-linked pattern of inheritance. To be affected, a female must have both of her X chromosomes carrying the recessive allele (a/a). Unaffected females may either have both X chromosomes carrying the dominant allele (A/A) or one dominant and one recessive allele (A/a). If any sons of an unaffected female are affected, then she must have the A/a genotype. There are only two male genotypes possible: affected males will have an X chromosome

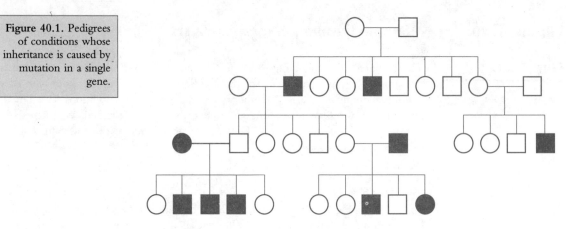

Figure 40.1. Pedigrees of conditions whose inheritance is caused by mutation in a single gene.

(a)

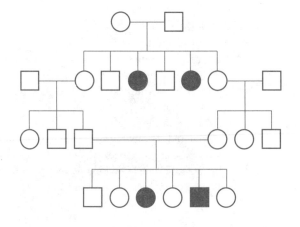

(b)

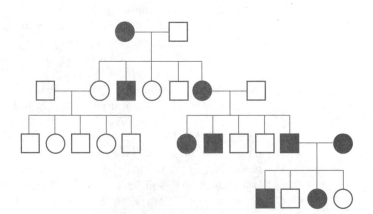

(c)

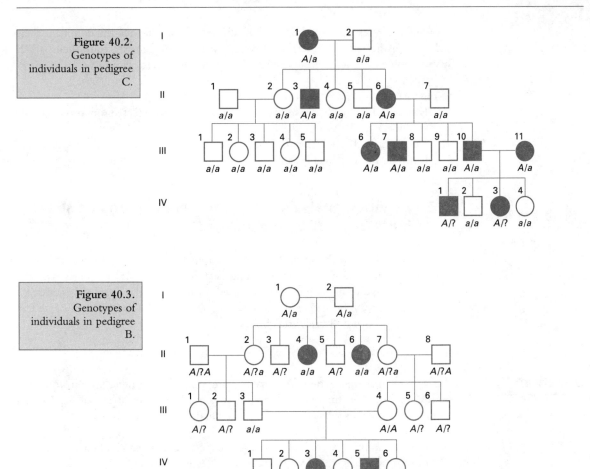

Figure 40.2.
Genotypes of
individuals in pedigree
C.

Figure 40.3.
Genotypes of
individuals in pedigree
B.

Figure 40.4.
Genotypes of
individuals in pedigree
A.

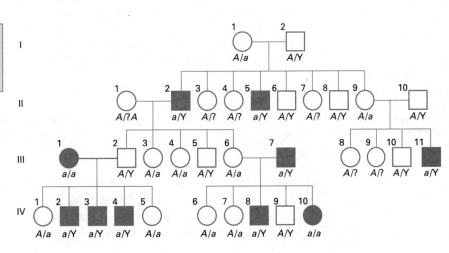

carrying the recessive allele (a/Y) and unaffected males will have the dominant allele (A/Y). Unaffected daughters of affected males must inherit from them an X chromosome carrying a recessive allele, and so will be heterozygous (A/a) (Fig. 40.4).

Note that:

1. Female II.1 may either be A/a or A/A. The absence of affected sons provides only very weak evidence that she is A/A.
2. Since female III.1 is affected, she must be homozygous for the recessive allele and hence all her sons will be affected and all her daughters heterozygous.

Problem 41	**Segregation analysis, simple risk estimation and the problem of ascertainment**

In human pedigree analysis, family sizes are too small to carry out statistical analysis of segregation ratios. However, by collecting data from a number of families with the particular condition, larger sample sizes from particular parental genotype combinations may be gathered and the correctness of the model may be tested. However, this is not entirely straightforward.

Question

Imagine 64 marriages where both parents are known to be heterozygous for cystic fibrosis, a condition that is caused by an autosomal recessive disease allele. In each family, let us suppose that there were three children.

1. In what number of families would you expect no affected children?
2. In what number would you expect one affected child?
3. In what number would you expect two affected children?
4. In what number would you expect three affected children?
5. If you were carrying out a survey of the ratio of affected children to unaffected children for cystic fibrosis, but you had no means of detecting carriers except by their production of affected children, which of the above types of family would you detect in your sample of cystic fibrosis families?
6. What would you expect for the ratio of unaffected to affected children in (1)–(4)?
7. What would the ratio be if families from (1) were excluded?

Answer

For an autosomal recessive condition, any child of two heterozygotes (carriers) has a probability of 0.25 (1/4 or 25%) of being homozygous recessive and therefore having the condition. The chance of an unaffected child is 0.75. In two children, the chance of them both being affected is 0.25×0.25, which is 0.0625 (or 1/16). The chance of having one affected and one unaffected is $2 \times 0.25 \times 0.75$, which is 0.375 (or 6/16), and the chance of having two unaffected children is 0.75×0.75 which is 0.5625 (or 9/16). (For one affected and the other unaffected, this is 0.25×0.75 for the first affected and the second unaffected plus 0.75×0.25 for the first unaffected and the second affected.) Therefore, for three children:

1. The probability of all three unaffected is $0.75 \times 0.75 \times 0.75$, which is 0.421875 (or 27/64). Thus 27 of our 64 families would be expected to have no affected children.
2. To have the first child affected and the next two unaffected, the probability is $0.25 \times 0.75 \times 0.75$ which is 0.140625 or 9/64. However, the birth order could be AUU, UAU or UUA where A is affected (not adenine) and U unaffected (not uracil)! Thus 27 of our 64 families would have just one affected child, in any position.
3. To have two affected children, the probability is $0.25 \times 0.25 \times 0.75$ which is 0.046875 or 3/64. However, there are three possible birth orders, AAU, AUA and UAA. Hence 9 of our 64 families would have two affected children.
4. To have all three children affected, the calculation is $0.25 \times 0.25 \times 0.25$ which 0.015625 or 1/64. Thus one of our 64 families would have all three children affected.
5. Summing the results from all 64 families:

	Unaffected	Affected
27 families with none affected	81	0
27 families with one affected	54	27
9 families with two affected	9	18
1 family with three affected	0	3
Total	144	48

This gives the expected 3 : 1 ratio for the offspring of two heterozygotes, showing that it is valid to sum offspring from different matings of the same genotype combinations.

6. If two carrier parent families could only be identified by having affected children, those marriages of two carriers that happen to produce no affected offspring would be missed and this would bias the calculation. This would alter the above ratio to 63 unaffected children and 48 affected, not a recognizable Mendelian ratio. This is the problem of ascertainment, caused by unintentional selection of data. Of course, the larger the family size, the more accurate ascertainment becomes, but for small families the error is very large. Depending on the methods used to find families, various forms of correction can be attempted. These are detailed in good human genetics textbooks, e.g. *Human Genetics: Problems and Approaches*, by Vogel and Motulsky (Springer-Verlag, 1979).

Problem 42 — Risk estimation from pedigrees

Question

People with a family history of a condition may ask for genetic counselling. Before considering the options open to them, they will need to know what the risk is that

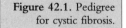

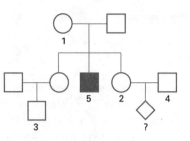

Figure 42.1. Pedigree for cystic fibrosis.

their next child will be affected by the condition. Imagine that you are a genetic counsellor and a couple with a family history of cystic fibrosis consult you; the woman is pregnant for the first time. After taking all the relevant details, you ascertain the pedigree in Fig. 42.1, in which individuals 2 and 4 are the couple consulting you.

Cystic fibrosis has an incidence in the UK of about 1 in 2000 births and as such it is the commonest autosomal recessive disorder in this country. You may assume for the time being that heterozygotes (carriers) for cystic fibrosis occur with a frequency of about 1 in 22, and new mutations are very rare.

1. Giving your reasoning, what are the approximate probabilities of individuals 1, 2 and 3 in Fig. 42.1 being heterozygous for the cystic fibrosis allele?
2. From this, and from the information above about the frequency of heterozygotes in the UK population at large, what is the probability that the first child of 2 and 4 will be affected with cystic fibrosis?
3. If the first child is unaffected but a second child has cystic fibrosis, what is the probability of a third child having the disease?
4. If, instead of the above scenario, neither of the first two children has cystic fibrosis, what is the chance that the third will have the disease?
5. Blood samples from individuals in this pedigree are taken and sent for screening with a DNA probe for cystic fibrosis allele ΔF_{508}, which is responsible for approximately 80% of all cystic fibrosis alleles in the UK. This probe will of course give negative results for any other cystic fibrosis allele, of which about 250 have been identified. Three possible results (i, ii and iii) of this screening test on individuals 2, 4 and 5 are given below. In each case, predict what would be the probability of the first child of individuals 2 and 4 being affected.

	Individual 2	Individual 4	Individual 5
Genotypes			
(i)	N/N	$N/\Delta F_{508}$	$\Delta F_{508}/\Delta F_{508}$
(ii)	$N/\Delta F_{508}$	N/N	$N/\Delta F_{508}$
(iii)	N/N	$N/\Delta F_{508}$	N/N

Answer

1. As individual 5 has cystic fibrosis (Fig. 42.1), both parents, being phenotypically normal, must be carriers. Hence individual 1 is a heterozygote. Individual 2, being the unaffected child of two carriers, has a 1/3 chance of being homozygous

normal and a 2/3 chance of being heterozygous. The unnumbered mother of individual 3 is genetically the same as her sister, individual 2. The father of individual 3 is unaffected and unrelated, and thus has a 1/22 chance of being a carrier. Individual 3 is unaffected and so must be either heterozygous or homozygous normal. To deduce his probable genotype, his father has a 1/22 chance of being a carrier, hence a 1/44 chance of contributing a gamete containing the disease allele. His mother has a 2/3 chance of being a carrier, hence a 1/3 chance of contributing a gamete with the disease allele. Hence, individual 3, who is not affected and hence does not result from the 1/132 chance of being homozygous for the disease allele, has 45/132 probability of being a carrier and 86/132 probability of being homozygous normal.

2. The woman seeking counselling (2) (Fig. 42.1) has the same carrier probability as her sister, and her unrelated husband the same carrier probability as the father of individual 3. Hence the probability of two affected gametes combining to produce the foetus is 1/44 from the father and 1/3 from the mother. Hence the probability of the foetus being affected is $1/44 \times 1/3 = 1/132$. Note that because cystic fibrosis is a very common condition, the disease allele is sufficiently common that the possibility of the father being a carrier has been included in the calculation. For much rarer alleles (like those of most genetic diseases), the remote possibility of an unaffected and unrelated individual being a carrier can be disregarded, unless there is evidence that the disease has occurred in that family also.

3. If the couple have a child with cystic fibrosis, then they are both heterozygous for the condition (carriers). Therefore, the probability must be recalculated and, as for any heterozygote by heterozygote cross for an autosomal recessive condition, the chances of the third child being affected would be 1/4.

4. If the first two children of the couple had been unaffected, the probability calculation would be no greater than for the first child (a 1/132 chance of an affected child). The fact that the first two were unaffected would reduce the probability that the father was a carrier, and thus of the third child being affected (to a probability of about 1/200).

5. (i) In this case, we can eliminate the possibility that the expectant mother is a carrier, as the disease allele for which her affected brother (5) is homozygous is ΔF_{508} and therefore this is the allele that both her parents carry. The fact that the unrelated father carries this allele is immaterial, as there is no chance of a homozygous affected offspring.

(ii) In this case, the affected brother (5) is homozygous for cystic fibrosis, but received a different allele from each carrier parent.* Female 2 received the ΔF_{508} allele from one parent, but as she is unaffected the other allele must be normal. Hence she is a carrier, so 1/2 of her gametes will carry a cystic fibrosis allele. Her spouse is unaffected, but may be a carrier. However, he does not carry ΔF_{508} so he has a reduced probability of being a carrier ($1/5 \times 1/22$) and therefore a

---

* Although having two disease alleles, as they are different disease alleles (only one being the ΔF_{508} allele), he is technically heterozygous. However, as both are disease alleles, he is homozygous for the disease.

probability of donating an affected gamete of $1/5 \times 1/22 \times (1/220)$. Combining these two affected gamete frequencies, the probability of the first child of 2 and 4 being affected is $1/440$.

(iii) Individual 5 is homozygous for two non-ΔF_{508} cystic fibrosis alleles, so his sister has a 2/3 chance of being a carrier and a 1/3 chance of being homozygous normal, hence a 1/3 probability of contributing an affected gamete. The unrelated 4 is heterozygous for ΔF_{508} but is phenotypically normal, so the other allele is normal; hence he has a 1/2 chance of contributing a disease gamete. Thus the probability of their first child having cystic fibrosis is $1/3 \times 1/2 = 1/6$.

Problem 43

Pedigree analysis of a disease gene and haplotype markers

The three pedigrees in Figs 43.1–43.3 show some hypervariable marker information for the autosomal dominant condition myotonic dystrophy. This is shown for eight

Figure 43.1. Pedigree showing hypervariable marker information for myotonic dystrophy: loci *AB*.

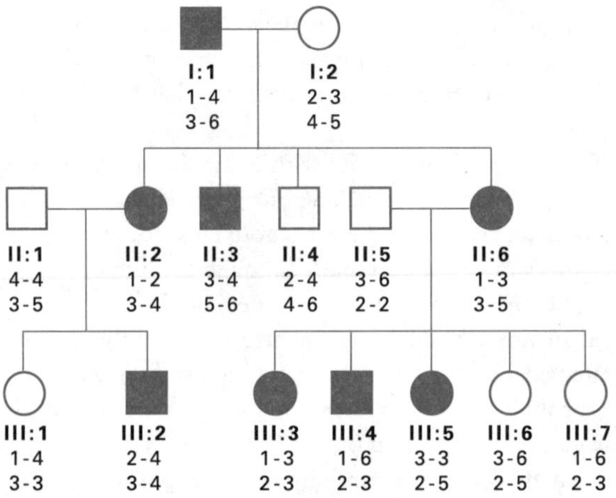

Figure 43.2. Pedigree showing hypervariable marker information for myotonic dystrophy: loci *CD*.

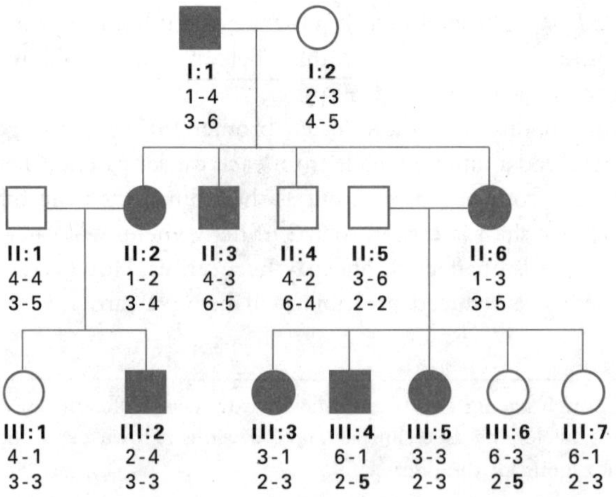

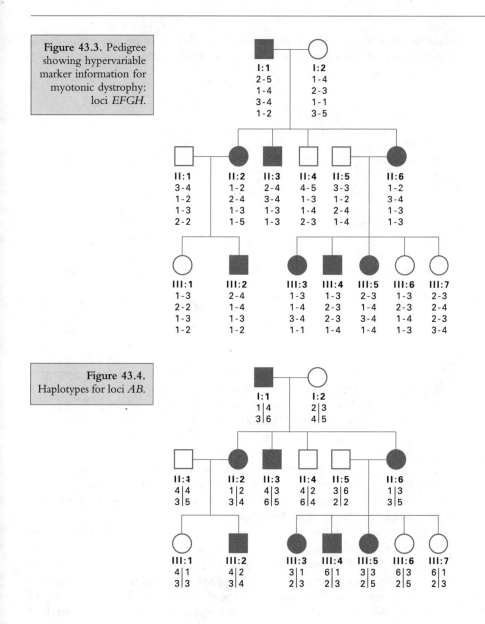

Figure 43.3. Pedigree showing hypervariable marker information for myotonic dystrophy: loci *EFGH*.

Figure 43.4. Haplotypes for loci *AB*.

loci, *A B* in Fig. 43.1, *C D* in Fig. 43.2 and *E F G H* in Fig. 43.3, all loci located on chromosome 19.

Under the symbol for each individual is a roman numeral indicating the generation and a unique arabic numeral indicating the individual, e.g. II:4 is individual 4 in the second generation. Below those numbers are haplotypes, which indicate the alleles at the relevant marker loci. Thus in Fig. 43.1, individual II:5 has the designation:

3-6
2-2

indicating that at locus A, one chromosome carries allele 3 and the other copy carries allele 6; at locus B, both copies of the chromosome carry allele 2.

Question

1. From Fig. 43.1, construct haplotypes for all individuals to show which allele of locus A is associated with which allele of locus B.
2. Is there any association between the haplotype and the disease?
3. For the marker loci C (upper) and D (lower) in Fig. 43.2, indicate haplotypes for all individuals.
4. Are there any individuals showing recombinant haplotypes? If so, which individuals are recombinant?
5. Is the disease locus linked to C and/or D?
6. For the marker loci E F G H in Fig. 43.3, also on chromosome 19, indicate haplotypes for all individuals.
7. The loci are linked in the order shown on the pedigree. Where and what recombinants are there?
8. What can you conclude about the location of the disease gene relative to loci E F G H?

Answer

1. To determine which parent donated which allele of which locus to the progeny is simple if, for example, at locus A one parent is of haplotype 1-3, the other of 2-4 and the offspring 1-4; it is obvious that allele 1 in the progeny comes from parent 1 and allele 4 from parent 2. If instead, parent 1 was 1-4 and parent 2 was 2-4, although both parents contain allele 4, as allele 1 must have come from parent 1, allele 4 in the progeny must in this case have come from the other parent.

   To associate alleles of the two loci, one must compare combinations of alleles found in the parents with those in the progeny. For haplotypes of linked genes, this allows one to determine the allele of each locus on each of the two chromosomes in the parental diploid genotype.

   The haplotypes for loci A B are given in Fig. 43.4, the numbers to the left of the vertical line indicating one chromosome type and those to the right representing the allele combination in the other chromosome. For generations II and III, with the exception of unrelated individuals marrying in, the numbers to the left represent the allele combination received from the left-hand parent and those to the right from the right-hand parent. No evidence is found of recombination between the marker loci.

2. For locus A, there are four alleles (1, 2, 3 and 4), with 1 and 4 found in the original affected individual. In his four offspring, three are affected and one unaffected. Allele 4 is found in one affected and one unaffected. Allele 4 is found in the unaffected son and two of the three affected children. Inv subsequent generations, female II:2 who inherited allele 1 from her affected father has an affected son, not carrying allele 1. Daughter II:6, who also inherited allele 1 from her affected father, has five children, three affected and two of those inheriting allele 1. Thus there is no statistical evidence of an association between the disease marker and any allele of locus A.

For locus *B*, male I:1 carries alleles 3 and 6, with two affected children inheriting 3 and one inheriting 6. The unaffected son II:4 also carries allele 6. Daughter II:2 passes on allele 3 to son III:2. Daughter II:6 passes on allele 3 to two of three affected children, and one of two unaffected. Again there is no significant evidence of linkage of the disease to locus *B*.

3. The haplotypes of individuals in Fig. 43.2 are shown in Fig. 43.5. The genotype of I:1 is based on combinations 1 3 and 2 4 being most frequent in generation II, while that of II:2 depends on the predominance of 2 4 and 3 5 combinations. II:1 is necessarily 4 3 and 4 5, as is II:5 (3 2 and 6 2).

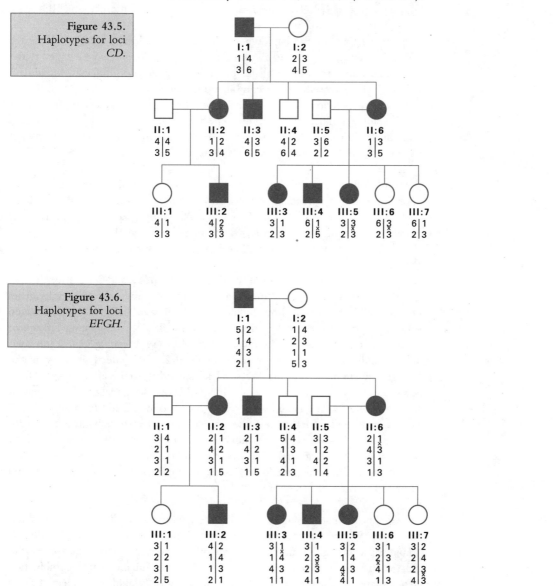

Figure 43.5. Haplotypes for loci *CD.*

Figure 43.6. Haplotypes for loci *EFGH.*

4. The recombinant haplotypes in generation III result from crossing over in meiosis in II:2 and II:6. Recombination has occurred between the two loci in meiosis in individuals II:2 and II:6, resulting in recombinant combinations in generation III individuals, marked with × in Fig. 43.5. All of the progeny of I:1 and I:2 have received alleles in the same combinations, so these are non-recombinant. Individual II:2 therefore has haplotype 1 3 for loci C D from her father, with the left one unaltered from father and 2 4 unaltered from mother. Grandson III:2 has received 4 3 from his father and recombinant 2 3 from mother. Female II:6 has received non-recombinant gametes from her parents: 1 3 and 3 5. Both III:5 and III:6 have received a recombinant 3 3 gamete from their mother, and III:4 has a recombinant 1 5 from her.

5. Individual I:1 carries the haplotypes 1 3 and 4 6 for loci C D. If the gene for myotonic dystrophy is on this chromosome, either haplotype may be linked to the autosomal dominant disease allele. In the next generation, affected individuals II:2 and II:6 carry the 1 3 combination, but II:3 carries the 4 6. In the succeeding generation, all progeny are the offspring of II:2 and II:6, both with the 1 3 haplotype. That non-recombinant haplotype is carried by one affected individual III:3 and two unaffected, III:1 and III:7. Among recombinant haplotypes, III:2 and III:6 carry allele 3 of locus D, and III:4 carries allele 1 of locus C.

    From analysis of the haplotypes and the occurrence of the disease, there is no clear correlation between any haplotype of either locus and the disease allele. Therefore, if there is any linkage between haplotype and disease allele, it is loose.

6. Haplotypes for the pedigree in Fig. 43.3 are given in Fig. 43.6.

7. Recombinants and the location of recombination events are indicated by × in Fig. 43.6.

8. The original affected male has the two haplotypes 5 1 4 2 and 2 4 3 1 for the loci E F G H. If they are linked to the autosomal dominant myotonic dystrophy gene, either combination may show linkage to the disease allele. All three affected offspring, but not the unaffected progeny, carry 2 4 3 1, suggesting linkage of these haplotype markers to the disease allele. In the succeeding generation, two affected individuals also carry an unrecombined 2 4 3 1, effectively proving the linkage of these markers to the disease allele. Individual III:4 is affected and is 1 3 3 1, so he only carries the same alleles of loci G H, suggesting that the disease gene is in the region of G and H and is not left of E or between E and F. Individual III:7 who is unaffected carries 2 4 3 3, the same as most affected individuals except for the haplotype 3 at locus H. This indicates that the recombination event between loci G and H, which gave this haplotype combination, removed the disease allele in that event. Thus, the gene for myotonic dystrophy is either between loci G and H or to the right of H.

We do know that all the hypervariable haplotype markers are located on chromosome 19. The fact that the gene for myotonic dystrophy is linked to loci G and H but shows no significant linkage to the other markers is not unexpected, as over longer genetic distances crossing over may be so frequent that recombination frequencies are not significantly different from what would be expected from independent assortment.

ADVANCED LINKAGE ANALYSIS

Introduction

Where a pair of genes are very tightly linked, it is often difficult to obtain sufficient recombinants to establish a reliable map by considering recombinants between the two genes in isolation. However, classification of the genotypes of other linked genes usually enables gene order to be established.

Linkage analysis can also be complicated by chromosome rearrangements. Translocations and inversions can mean that two strains which are crossed may not give the same recombination frequencies or even have the same gene order. Recombination within the region of the chromosome involving such rearrangements usually gives rise to inviable progeny. Why?

Other complications not directly connected with linkage arise because the genes involved may have phenotypes that interact, i.e. epistasy (the masking of one phenotype by the phenotype associated with the genotype caused by a different gene, e.g. the inability to score eye colour in individuals without heads). If epistasy is present, then it may be necessary to confine the linkage analysis to only a sub-set of the test-cross progeny. It is important that this sub-set is balanced, i.e. excludes equal numbers of recombinant and parental classes (but not individuals).

Problem 44 — Complex linkage analysis with rearrangements

Question

In a cross between a monster from a stock true-breeding for the absence of heads with a monster from a stock true-breeding for blue eyes, all the F_1 had heads with green eyes. The F_1 were crossed to a true-breeding headless stock that is known as a result of progeny testing to be homozygous for blue eyes. One hundred test-cross progeny were classified as follows:

headless	50
green-eyed	45
blue-eyed	5

The test-cross can be represented symbolically as:

$$\frac{hdl \; +}{+ \; blu} \times \frac{hdl \; blu}{hdl \; blu}$$

The test-cross progeny are:

$$\frac{hdl \ +}{hdl \ blu} \qquad \frac{+ \ blu}{hdl \ blu} \qquad \frac{hdl \ blu}{hdl \ blu} \qquad \frac{+ \ +}{hdl \ blu}$$

headless   blue-eyed   headless   green-eyed

The proportion of green-eyed progeny among all progeny with heads is a valid estimate of the recombination fraction (i.e. $5/(45 + 5) = 0.1$ or 10%).

This data set relates to genes on one chromosome of this monster, the tissues of which are diploid, as you will have concluded already from the data given above. The data have also provided an estimate of the map distance between the eye colour and head presence gene.

1. You have available two true-breeding stocks, one clawless but having heads, the other having claws but without heads. Clawless is recessive to clawed, headless is recessive to headed. Explain how you would determine the recombination frequency between the genes for claw and head presence/absence.
2. The F_1 from a cross between a true-breeding headless, clawless, dwarf stock and a true-breeding tall stock having heads and claws was tall and had both heads and claws. F_1 individuals were crossed to individuals from the true-breeding headless, clawless, dwarf stock to give progeny as shown in Table 44.1. Map the genes concerned in these crosses.

Table 44.1. Progeny of test-cross segregating for head, claw and height.

Phenotype	Frequency
headless, clawless, dwarf	169
headless, clawless, tall	21
headless, clawed, dwarf	8
headless, clawed, tall	1
headed, clawless, dwarf	1
headed, clawless, tall	11
headed, clawed, dwarf	18
headed, clawed, tall	175

Table 44.2. Test-cross segregating for height, claws, eye colour and tail shape.

Phenotype	Frequency
dwarf, clawless, green-eyed, forked tail	53
dwarf, clawless, green-eyed, simple tail	152
dwarf, clawed, green-eyed, forked tail	10
dwarf, clawed, green-eyed, simple tail	26
dwarf, clawed, blue-eyed, simple tail	2
tall, clawless, green-eyed, forked tail	2
tall, clawless, blue-eyed, forked tail	31
tall, clawless, blue-eyed, simple tail	7
tall, clawed, blue-eyed, forked tail	163
tall, clawed, blue-eyed, simple tail	46

Phenotype	Frequency
clawless, forked tail	75
clawless, simple tail	32
short claws, forked tail	23
short claws, simple tail	2
long claws, forked tail	7
long claws, simple tail	69

Table 44.3. Test-cross progeny segregating for claws, claw length and tail shape.

Phenotype	Frequency
clawless, one blue eye, forked tail	85
clawless, two blue eyes, forked tail	9
clawless, one blue eye, simple tail	1
clawless, two blue eyes, simple tail	10
clawed, one green eye, forked tail	10
clawed, two green eyes, forked tail	1
clawed, one green eye, simple tail	13
clawed, two green eyes, simple tail	89

Table 44.4. Test-cross segregation of claws, eye colour, eye number and tail shape.

3. A monster from a clawless, dwarf true-breeding stock is crossed to one from a stock that breeds true for blue eyes and a forked tail. All the F_1 are tall and have claws, green eyes and a simple tail. F_1 individuals are crossed to individuals from a true-breeding dwarf stock that is clawless, has blue eyes and forked tails. Progeny are obtained as shown in Table 44.2. Use these data to extend your chromosome map.

4. F_1 monsters were obtained by crossing a true-breeding clawless stock with forked tails to true-breeding clawed stock with simple tails. The F_1, which had claws and simple tails, were crossed to individuals from the true-breeding clawless stock with forked tails. The progeny shown in Table 44.3 were obtained. Having noticed the different claw morphologies among the test-cross progeny, the F_1 and clawed parental stock were rescored and both were seen to have long claws. Add the gene for claw length to the chromosome map.

5. A new monster stock from a different planet (but of the same species) has been obtained. This has two eyes, unlike all the stocks previously studied, which have only one. The new stock is crossed to a true-breeding clawless, blue-eyed, forked-tailed stock. The F_1 are uniform, all having claws, two green eyes and simple tails. The F_1 are crossed to the true-breeding clawless, blue-eyed, forked-tailed stock and the progeny classified in Table 44.4 are obtained. Analyse these results and comment on how they relate to the map that you have already established.

Answer

1. Cross stocks together to obtain the F_1, which will be clawed and have heads. The F_1 must be crossed to a true-breeding clawless, headless stock (the double homozygous recessive genotype). If this stock is not available, it could be obtained by first breeding an F_2 (NB the double homozygous recessive genotype will have a unique phenotype, but may be quite rare depending on how tightly the genes are linked). The test-cross will allow the recombination frequency to be estimated as the percentage of recombinant progeny (clawless, headless and clawed, headed).

2. The total number of progeny obtained from the cross is 404. These are classified in Table 44.5. From these estimates of inter-gene distances, the linkage map shown in Fig. 44.1 can be derived.

3. The total number of progeny scored in this cross was 492 and their classification is shown in Table 44.6. These data allow us to add to the map in Fig. 44.1 to give the extended linkage map shown in Fig. 44.2. Note that the relative order of *blu*

Gene pair	Number of recombinants	Recombination fraction (%)
headed/headless and clawed/clawless	$8 + 1 + 1 + 11 = 21$	5.2
headed/headless and tall/dwarf	$21 + 1 + 1 + 18 = 41$	10.1
clawed/clawless and tall/dwarf	$21 + 8 + 11 + 18 = 58$	14.4

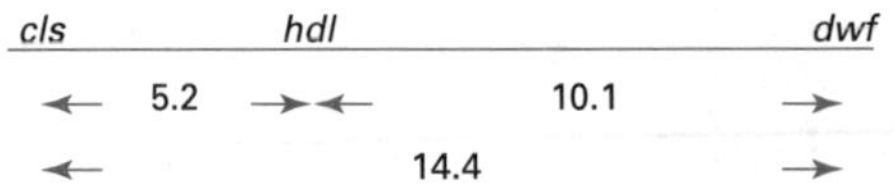

Table 44.5. Progeny from the test-cross.

Figure 44.1. Linkage map of the head, claw and height genes.

Gene pair	Number of recombinants	Recombination fraction (%)
clawed/clawless and tall/dwarf	$10 + 26 + 2 + 2 + 31 + 7 = 78$	15.9
clawed/clawless and green/blue	$10 + 26 + 31 + 7 = 74$	15.0
clawed/clawless and simple/forked	$53 + 26 + 2 + 2 + 31 + 46 = 160$	32.5
tall/dwarf and green/blue	$2 + 2 = 4$	0.8
tall/dwarf and simple/forked	$53 + 10 + 7 + 46 = 116$	23.6
green/blue and simple/forked	$53 + 10 + 2 + 2 + 7 + 46 = 120$	24.4

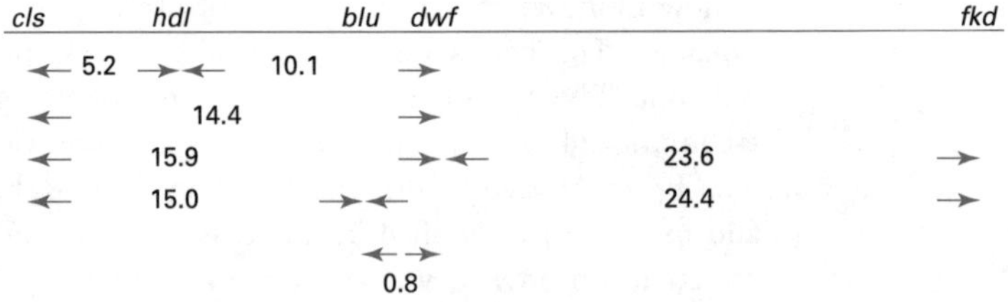

Table 44.6. Classification of the claw, head, tail, height and eye colour genes.

Figure 44.2. Extended linkage map.

Table 44.7. Classification of progeny for claw possession and length and for tail type.	Gene pair	Number of recombinants	Recombination fraction (%)
	clawed/clawless and simple/forked	$32 + 23 + 7 \doteq 62$ (208)	29.8
	clawed/clawless and long/short	$23 + 2 = 25$ (101)	24.8
	simple/forked and long/short	$2 + 7 = 9$ (/101)	8.9

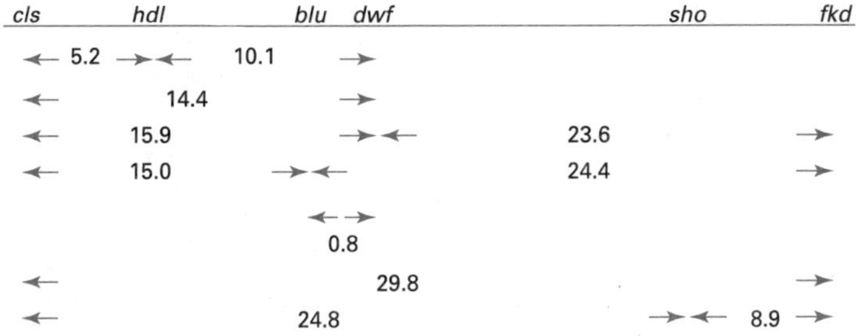

Figure 44.3. Extended linkage map of six genes.

and *dwf* can be confirmed by examining the four recombinants between these two loci. These are generated, using this map order, without further recombination.

4. The total number of progeny is 208, but claw morphology can only be scored in 101 of these progeny. The classification of progeny is shown in Table 44.7. Adding these data gives the extended map shown in Fig. 44.3.

5. For the final cross, the total number of progeny is 218 and they are classified in Table 44.8. Note first that there is no recombination between clawed/clawless and green/blue (which in previous crosses gave a recombination frequency of 15.0%). Analysis of these data is for green/blue, but identical values will therefore apply for clawed/clawless. This gives the map shown in Fig. 44.4. This is obviously different from the map progressively established in Figs 44.1–44.3. The cause is an inversion with one breakpoint somewhere between *blu* and *fkd* (recombination in this interval reduced from 24.4% to 10.1%) and the other beyond and to the left of *cls* (recombination suppressed between *cls* and *blu*). Inversions are loosely termed 'crossover suppressors' by geneticists. In fact, the major cause is not suppression of crossing over in the inverted region but the non-viability of such recombinant products due to the generation of unbalanced

Table 44.8. Analysis of the cross involving eye colour, eye number and tail shape.	Gene pair	Number of recombinants	Recombination fraction (%)
	green/blue and simple/forked	$1 + 10 + 10 + 1 = 22$	10.1
	green/blue and two/one eyes	$9 + 10 + 10 + 13 = 42$	19.3
	simple/forked and two/one eyes	$9 + 1 + 1 + 13 = 24$	11.0

genomes (duplication of the chromosome part to one side of the inversion and deletion of the part to the other side). Hence crossing over in the inversion is lethal, so recombinant products do not appear in the (live) progeny.

Figure 44.4.
Anomalous linkage map for three loci.

GLOSSARY

This glossary relates not just to the material covered in this book but to the subject of genetics as a whole, and as such should serve as a useful reference.

Acentric chromosome	A chromosome that results directly or indirectly from a chromosomal rearrangement, that has no centromere and will hence be lost in cell division.
Acrocentric chromosome	A chromosome with the centromere near one end, so that one arm of the chromosome is short and the other is long.
Adaptation	A characteristic of an organism that allows it to cope better with its environment; the evolutionary process by which organisms become adapted to their environment.
Additive gene	Gene affecting a given character whose alleles show no dominance to each other and no epistasis to other genes similarly affecting the same character.
Allele	One of a number of alternative forms of a gene, each possessing a unique nucleotide sequence; different alleles of a given gene are usually recognized, however, by the phenotypes rather than by comparison of their nucleotide sequences.
Allele frequency	A measure of the frequency of an allele in the population, i.e. the proportion of all alleles of the gene in question which are of the particular type.
Allopatric	Of populations or species that inhabit separate geographic regions (cf. **sympatric**).
Allopolyploidy	Polyploidy due to the addition of chromosome sets from two different species.
Allosteric transition	A change from one conformation of a protein to another conformation.
Allozymes	Alternative enzyme forms encoded by different alleles at the same locus.
Amber codon	One of the three stop codons, UAG, in the genetic code; a nonsense codon.
Amber mutant	A chain termination mutant caused by the change of a sense codon to the UAG stop codon.
Amino acids	The building blocks of proteins; several hundred are known, but only 20 are normally found in proteins.
Aminoacyl-tRNA	A tRNA molecule covalently bound to an amino acid via an acyl bond between the carboxyl group of the amino acid and the 3′-hydroxyl of the tRNA.
Amino terminus	The end of a polypeptide with the free amino group on the terminal amino acid residue. Unless subsequently removed by post-translational processing, this residue is normally methionine or *N*-formyl-methionine.
Amniocentesis	A procedure for sampling the amniotic contents, which can be used for diagnosing genetic abnormalities in a foetus.
Anabolic	Pertaining to an enzymatic reaction leading to the synthesis of a more complex biological molecule from a less complex one (cf. **catabolic**).
Anaphase	The third stage of mitosis or of meiosis (I or II) during which the chromosomes are pulled toward opposite poles of the cell.
Aneuploid	A cell with a chromosome number deviating from the basic haploid number or a multiple thereof.
Annealing	The process, also called nucleic acid hybridization, by which two single-stranded polynucleotides form a double-stranded molecule, with hydrogen bonding between the complementary nucleotides of the two strands. Annealing can take place between

	complementary strands of either DNA or RNA to produce double-stranded DNA molecules, double-stranded RNA molecules or RNA–DNA hybrid molecules.
Antibody	A protein (an immunoglobulin), synthesized by the immune system of a higher organism, that binds specifically to the foreign molecule (**antigen**) that induced its synthesis.
Anticodon	The three adjacent nucleotides in a tRNA molecule that are complementary to, and pair with, the three nucleotides of a codon in an mRNA molecule during protein synthesis.
Antigen	Any substance capable of stimulating production of specific antibodies. See **antibody**.
Antiparallel	In DNA, the two strands in a double helix have opposite molecular orientations with respect to the phosphodiester backbone.
Apoptosis	See **programmed cell death**.
Artificial selection	The process of choosing the parents of the following generation on the basis of one or more genetic traits (cf. **natural selection**).
Ascus	A sac containing **ascospores**, the products of a single meiosis, in the fruiting bodies of ascomycete fungi.
Asexual reproduction	The development of an organism from one or more cells in the absence of any sexual process; also called **vegetative reproduction**.
Assortative mating	Non-random selection of mates with respect to one or more traits. Negative assortative mating is the preferential selection of mates with different genotypes, positive with similar.
ATP	Adenosine triphosphate, a primary repository of chemical energy in cells.
Autogamy	The condition found in ciliate protozoans for example, in which a cell with a single diploid nucleus can undergo sexual reproduction.
Autopolyploidy	Polyploidy due to the presence of more than two chromosome sets of the same species.
Autoradiography	A technique by which molecules or structures labelled by incorporation of radioisotopes can be visualized by a photographic technique.
Autosome	A chromosome other than a **sex chromosome**.
Auxotroph	A strain of organisms unable to synthesize a given molecule required for their own growth; growth can occur when the required compound is supplied in the food (cf. **prototroph**).
Back-cross	The cross of an organism with one of its parents, or with an organism of identical genotype thereto.
Back mutation	A mutation that causes a mutant allele of a gene to regain, completely or partially, its wild-type function (= **reversion**).
Bacteriophage	A virus whose host is a bacterial cell; also called **phage**.
Balanced lethals	Recessive lethals at different loci, so that each homologous chromosome carries at least one lethal, and associated with inversions, so that no viable recombinant results from crossing over between the homologous chromosomes.
Base analogue	A purine or pyrimidine base other than those normally found in DNA or RNA, which may be incorporated into DNA or RNA if supplied at the time of replication or transcription. Incorporation of a base analogue may result in mutation.
Base composition	The overall proportion of the different bases in the DNA of the genome.
Base pair	The nitrogenous bases that pair by hydrogen bonding in double-stranded DNA or RNA.
Base substitution	The substitution of a base in DNA by another different base.

Bi–directional replication	The replication of DNA in both directions on both strands, from a replication origin.
Binding site	The target sequence in DNA or RNA to which a specific (regulatory) protein binds.
Bivalent	Two homologous chromosomes pair during the early stages of first meiotic division.
Blastocyst	The embryonic stage of mammals that consists of about 64 cells.
Blastoderm	A multinuclear embryonic stage that results from nuclear divisions, without division of the cytoplasm of the zygotic cell.
Blastula	A multicellular embryonic stage resulting from complete cleavage divisions of the zygotic cell into a number of smaller cells.
Blunt end ligation	Ligation of the blunt ends of two DNA molecules, the ends usually having been produced by cleavage with restriction enzyme(s) leaving blunt ends from a transverse cleavage, as opposed to complementary single-stranded ends from an offset cut.
Bottleneck	A period when a population becomes reduced to only a few individuals.
Branch migration	The migration of a half-chiasma in an intermediate stage of recombination, leading to the spread of heteroduplex.
Carboxy terminus	The end of a polypeptide at which the terminal amino acid residue presents a free carboxyl group; normally corresponding to the last codon in the open reading frame, immediately adjacent to the chain termination codon.
Catabolic	An enzymatic reaction leading to the breakdown of a complex biological molecule into less complex components, which may either yield energy (e.g. in the form of ATP) or be used in subsequent anabolic reactions.
cDNA	Complementary DNA made *in vitro* by the enzyme **reverse transcriptase** on a template of RNA, usually mRNA.
Cell cycle	The growth cycle of an individual cell. The cycle of events between the formation of a new cell and its eventual division.
Centromere	A chromosomal region that becomes associated with the spindle fibres during mitosis and meiosis; also called **kinetochore**.
Chiasmata	The places at which pairs of homologous chromatids are in contact (from late prophase of meiosis to the beginning of first anaphase) and at which exchanges of homologous parts between non-sister chromatids have taken place by crossing over (singular: chiasma).
Chromatid	In all duplicated chromosomes each of the two longitudinal subunits that become visible during mitosis or meiosis up to the time of their separation at anaphase.
Chromatin	Material containing DNA, histone proteins, non-histone proteins and some RNA, identifiable in the nuclei of cells by its special staining properties.
Chromomere	A bead-like region in partially condensed chromatin.
Chromosomal aberration	A gross change in the structure of a chromosome, adding to, removing from, or altering its sequence, e.g. inversion, translocation, duplication, deletion.
Chromosomal polymorphism	The presence in a population of more than one gene sequence for a given chromosome.
Chromosome	A thread-like structure, found in the nuclei of cells, that contains the genes arranged in linear sequence; a whole DNA molecule comprising the genome of a prokaryotic cell; a DNA molecule complexed with histones and other proteins in eukaryotic cells.
Chromosome set	The normal gametic (i.e. haploid) complement of chromosomes of an individual.
Cis–trans test	A genetic test for allelism. See **complementation test**.
Cistron	A nucleotide sequence in DNA specifying a single genetic function as defined by the complementation test; a nucleotide sequence coding for a single polypeptide; a gene.

Cline	A gradient in the frequencies of alleles, and hence genotypes or phenotypes along a stretch of territory.
Clone	(The process of isolating) a group of genetically identical organisms, cells or nucleic acid sequences.
Coadaptation	The harmonious interaction of certain alleles of different genes; the selection process by which harmoniously interacting combinations become established in a population.
Coding strand	The (sense) strand of the DNA double helix that has the same polarity and sense as the mRNA transcribed. The other (antisense) strand is the actual template for the transcribed RNA.
Codominance	The situation in a heterozygote in which both alleles are expressed in the phenotype (e.g. the human A and B blood group alleles).
Codon	A group of three adjacent nucleotides in an mRNA molecule that code either for a specific amino acid or for polypeptide chain termination during protein synthesis.
Coefficient of selection	The intensity of selection, as measured by the proportional reduction in the gametic contribution of one genotype compared with another.
Colinearity	The linear correspondence between the order of amino acids in a polypeptide chain and the corresponding nucleotide sequence in the DNA molecule, or between the order of genes on a recombinationally determined linkage map of a chromosome and the physical map of the chromosome.
Complementary DNA	See **cDNA**.
Complementation group	A set of mutant alleles, each recessive to wild type, exhibiting mutant phenotypic behaviour in the heterozygous state with each other.
Complementation test	A genetic test to ascertain whether two gene mutations occur in the same functional gene and to establish the limits of the functional gene; also called *cis–trans* **test**.
Conditional lethal mutant	A mutation that kills the affected organism under one set of environmental conditions (**restrictive condition**) but that is not lethal under another set of conditions (**permissive condition**).
Conditional mutant	A mutant that is only phenotypically different from the wild type under certain conditions, e.g. of temperature or pH. See **conditional lethal mutant**.
Conidium	The vegetative spore of an ascomycete fungus.
Conjugation	The process by which genetic material is transferred from one organism to another during cell-to-cell contact.
Conjugative plasmid	A bacterial plasmid capable of bringing about conjugation (usually resulting in transfer of its own DNA only).
Consanguinity	The sharing of at least one recent common ancestor.
Contig	Overlapping cloned fragments of DNA.
Continuous variation	Variation, with respect to a certain trait, among phenotypes that cannot be classified into a few clearly distinct classes.
Controlling element	A eukaryotic transposable element, detectable through the abnormal activity of standard genes that it affects.
Crossing over	The exchange of chromatid segments between homologous chromatids during meiosis; if different alleles are present at different loci on the chromatids, crossing over can be detected by the formation of genetically recombinant progeny.
Darwinian fitness	The relative fitness of one genotype compared with another, as determined by its relative contribution to the following generations.
Degenerate code	A code in which a single element in one language is specified by more than one element in the second language, e.g. the single amino acid isoleucine is specified by

	three different codons, i.e. in the universal genetic code, 20 amino acids are encoded by 61 different codons.
Deletion	A chromosomal mutation characterized by the loss of a chromosome segment (cf. **duplication**).
Deletion mapping	Mapping of sites within a gene or chromosome by the use of overlapping deletions.
Denatured DNA	DNA that has been converted from double-stranded to single-stranded form by breaking the hydrogen bonds joining the two complementary strands, i.e. by melting (cf. **native DNA**).
Denatured protein	A protein that has lost its natural conformation by exposure to a destabilizing agent such as heat.
Deoxyribonucleic acid (DNA)	A polynucleotide in which the sugar residue is deoxyribose and which is the primary genetic material of all cells.
Derepression	Gene activation by the inactivation of an otherwise active repressor.
Diakinesis	A stage in meiotic prophase I.
Dihybrid cross	A cross between individuals that have different alleles at each of two gene loci.
Dikaryon	A single cell with two genetically different nuclei.
Dioecious	Of organisms (usually plants) having either male or female sex organs but not both (cf. **monoecious**).
Diploid	A cell, tissue or organism having two chromosome sets (cf. **haploid, polyploid, tetraploid, triploid**).
Diplotene	A stage in meiotic prophase I.
Dizygotic twins	Twins developed from two separate fertilized ova; also called **fraternal twins**.
DNA	See **deoxyribonucleic acid**.
DNA gyrase	See **topoisomerase**.
DNA ligase	An enzyme that creates a phosphodiester bond between the 5'-phosphate end of one polynucleotide and the 3'-hydroxyl end of another, thereby producing a single, larger polynucleotide; also called **polynucleotide ligase**.
DNA polymerase	The enzyme responsible for synthesizing DNA from deoxyribonucleoside triphosphates under the direction of a template DNA strand; several DNA polymerase genes and enzymes with somewhat different properties and functions exist in the genome.
Dominance	A relationship between two alleles, where if one is dominant it is expressed in the phenotype of the heterozygote.
Down's syndrome	A syndrome characterized by physiological, behavioural and mental defects that are due to the presence of an extra copy of the genetic material contained in chromosome 21, normally as a free additional copy of chromosome 21 but occasionally translocated onto another chromosome.
Drift	See **random genetic drift**.
Duplex DNA	DNA in its native, hydrogen-bonded, double helical form.
Duplication	A chromosomal mutation characterized by the presence of two copies of a chromosome segment in the haploid genome.
Effective population size	Average number of reproducing individuals in a population.
Effector molecule	A (generally) small molecule whose concentration regulates the activity of a protein molecule by interacting with a specific binding site on the protein, causing an allosteric transition to occur in the structure and function of the protein molecule.
Electrophoresis	A technique for separating molecules based on their differential mobility in an electric field.

Endonuclease	An enzyme that hydrolyses internal phosphodiester bonds in a polynucleotide (cf. **exonuclease**).
Endosperm	In flowering plants, a tissue specialized for nourishing the developing embryo.
Episome	A genetic element (DNA molecule) that may exist either as an integrated part of a chromosomal DNA molecule of the host or as an independently replicating DNA molecule (plasmid or virus) free of the host chromosome.
Epistasis	Interaction between two different genes so that an allele of one of them (**epistatic** gene) interferes with, or even inhibits, the phenotypic expression of the other (**hypostatic** gene).
Euchromatin	Chromosome regions or whole chromosomes that have normal staining properties and that undergo the normal cycle of chromosome coiling, in contrast to the differentially condensing and staining **heterochromatin**.
Eugenics	The study of methods under social control directed toward 'improving' the hereditary constitution of future human generations.
Eukaryote	An organism that has a nucleus, i.e. all organisms except bacteria and blue-green algae.
Euploid	A cell, tissue or organism that has one or more exact multiples of a chromosome set (cf. **aneuploid**).
Exon	The DNA of a eukaryotic transcription unit whose transcript becomes a part of the mRNA produced by splicing out introns or parts of leader sequence from an hnRNA molecule.
Exonuclease	An enzyme that hydrolyses terminal phosphodiester bonds (at $3'$ or $5'$ ends) in a polynucleotide (cf. **endonuclease**).
Expressivity	The degree of phenotypic expression of a genotype.
Fertility factor	A conjugative plasmid capable of bringing about the transfer of chromosomal fragments by conjugation.
Fertilization	The fusion of two gametes to form a zygote.
Fitness	The reproductive contribution of an organism or genotype to the following generations (cf. **Darwinian fitness**).
Forward mutation	A mutation from the 'wild' type to the mutant condition (cf. **back mutation**).
Founder effect	Genetic drift due to the founding of a population by a small number of individuals whose allele frequencies at certain genes may not reflect those of the population as a whole.
Frameshift	A mutation caused by either the insertion or deletion in DNA of a number of nucleotide pairs, not divisible by three, whose effect is to change the reading frame of codons in an mRNA molecule during protein synthesis, causing an abnormal amino acid sequence to be synthesized from the site of the mutation on.
Fraternal twins	See **dizygotic twins**.
Gamete	A mature reproductive cell capable of fusing with another to give a **zygote**; also called **sex cell**.
Gametophyte	The haploid sexual generation that produces the gametes in plants in which a sexual (haploid) generation alternates with an asexual (diploid) generation (cf. **sporophyte**).
Gastrula	The embryonic stage marked by the beginning of cell movement and the initiation of organogenesis.
Gene	In the genome of an organism, a sequence of nucleotides to which a specific function can be assigned (e.g. a nucleotide sequence coding for a polypeptide, or a nucleotide sequence specifying a tRNA), defined at the phenotypic level by differences between the expression of different alleles of the same gene.

Gene amplification	*In vivo* extra replication of certain parts of the genome, generally to enhance expression.
Gene expression	The process by which the information content of a gene is normally transcribed, translated and possibly further modified to result in a functional manifestation of the encoded information.
Gene flow	The exchange of alleles (in one or both directions) at a low rate between two populations, due to the dispersal of gametes or of individuals from one population to another; also called gene **migration**.
Gene frequency	A jargon term, sometimes used in population genetics, for referring to the frequency of a particular allele of a gene. Hence the correct, unambiguous term is **allele frequency**.
Gene pool	The sum total of the genetic information in a breeding population.
Genetic variance	The fraction of the phenotypic variance that is due to differences in the genetic constitution of individuals in a particular population in a particular environmental range.
Genome	The genetic content of a cell, plasmid or virus; for diploids there is some ambiguity, as either the entire diploid chromosome set or just one (haploid) set may be so described.
Genotype	The sum total of the genetic information contained in an organism; the genetic constitution of an organism with respect to one or a few gene loci under consideration (cf. **phenotype**).
Germ cell	An animal cell set aside early in embryogenesis that may multiply by mitosis or produce, by meiosis, cells that develop into either eggs or sperm (cf. **somatic cells**).
Haploid	A cell, tissue or organism having one chromosome set.
Haplotype	The genetic constitution of a single chromosome or region of a chromosome; particularly used in the context of characterizing the major human incompatibility complex (HLA).
Heterochromatin	See **euchromatin**.
Heteroduplex	A double-stranded DNA molecule in which one strand was not synthesized on the other as template; instead, they originate from different but homologous double helices.
Heterosis	The superiority of a heterozygote over either homozygote for a particular trait.
HnRNA	The primary eukaryotic (heterogeneous nuclear) RNA transcript, prior to processing to remove ends of the transcript and introns in the maturation of mRNA.
Homeobox	A feature of homeotic genes; a sequence of about 180 nucleotides, encoding an approximately 60-amino acid motif in the translated protein which binds to DNA.
Homeotic gene	A key regulatory gene in development, specifying the development of higher order tissues and structures.
Homologous	Of chromosomes or chromosome segments that are normally identical with respect to their constituent genetic loci and their visible structure (exceptions arise from structural mutation, e.g. deletion), but with probable minor variation in base sequence; in evolution, of genes and structures that are similar in different organisms owing to their having inherited them from a common ancestor.
Homologous recombination	Genetic exchange between identical or nearly identical DNA sequences.
Homozygosity	The condition of being homozygous; the proportion of individuals homozygous at a locus or of homozygous loci in an individual.

Homozygote	A cell or organism having identical alleles at a given locus on homologous chromosomes.
Hybrid·	An offspring of a cross between two genetically unlike individuals.
Hybrid inviability	Reduction of somatic vigour or survival rate in hybrid organisms.
Hybridoma	A somatic cell hybrid between a tumour cell and an antibody-producing B cell.
Hybrid vigour	See **heterosis**.
Hypha	The filamentous cellular structure composing the body of a fungus.
Hypostasis	See **epistasis**.
Inbreeding	Mating between relatives.
Inbreeding coefficient	The probability that the two alleles at a locus in a diploid organism are identical by descent.
Inbreeding depression	Reduction in fitness or vigour due to the inbreeding of normally outbreeding organisms.
Incomplete dominance	In a heterozygote, where the phenotype is intermediate between that of either homozygote, but usually closer to that of the homozygote for the 'dominant' allele.
Incomplete penetrance	A mutant genotype in which not all individuals of that genotype express the mutant phenotype.
Inducer	An effector molecule responsible for the induction of enzyme synthesis.
Induction	(1) The synthesis of new enzyme molecules in response to an environmental stimulus; (2) the stimulation of lysogenic bacteriophage to switch to the lytic cycle.
Insertion sequence	One of a number of different nucleotide sequences found in cells and capable of moving from one chromosomal location to another; also known as **IS elements**.
Intercalation	Insertion into a DNA double helix, between adjacent bases, of a non-base planar molecule.
Interference	A measure of the degree to which one crossover by a chromatid affects the probability of a second crossover by that same chromatid.
Interphase	The state of the cell cycle during which metabolism and synthesis occur between cell divisions.
Intersex	An individual of a normally dioecious species whose reproductive organs or secondary sex characters are partly of one sex and partly of the other.
Intervening sequence	See **intron**.
Intron	A non-coding nucleotide sequence in eukaryotic DNA and hnRNA, separating two portions of nucleotide sequence found to be contiguous in cytoplasmic mRNA.
Inversion	A chromosomal mutation characterized by the reversal of a chromosome segment within a chromosome.
Inversion polymorphism	The presence of two or more chromosome sequences, differing by inversions, in the homologous chromosomes of a population.
Inverted repeat	Two copies of a DNA sequence found in opposite orientation in the same DNA molecule, characteristic of the ends of **transposons**.
In vitro	(An experiment) carried out with isolated cells from a multicellular organism with disrupted cells or with sub-cellular fractions thereof.
In vivo	(An experiment) carried out within an organism.
IS element	See **insertion sequence**.
Isolating mechanism	See **reproductive isolating mechanism**.
Karyotype	The chromosome complement of a cell or organism, characterized by the number, size and configuration of the chromosomes.
Kinetochore	See **centromere**.

Klinefelter's syndrome	A human syndrome that is due to the presence of one extra X chromosome in the male karyotype (XXY).
Lagging strand	In DNA replication, the strand synthesized by DNA polymerase discontinuously as a series of Okazaki framents (cf. **leading strand**).
Leader	The region of an mRNA molecule extending from the 5' end to the beginning of the coding region of the first structural gene; it may contain ribosomal binding sites for example.
Leading strand	In DNA replication, the strand synthesized continuously by DNA polymerase (cf. **lagging strand**).
Leptotene	A stage in meiotic prophase I.
Lethal	An allele that causes death (in all carriers if it is dominant, but only in homozygotes if it is recessive) before reproductive age.
Ligase	See **DNA ligase**.
Linkage	A measure of the degree to which two different genes assort by crossing over at meiosis or in genetic crosses.
Linkage disequilibrium	Non-random association of alleles at different loci in a population.
Linkage group	A set of genes that can be placed in a linear order representing the different degrees of linkage between the loci.
Linkage map	A map of the chromosome or chromosomes of a species based on recombination frequency data from crosses.
Locus	The place at which a particular mutation or a gene resides in a genetic map; often used interchangeably with gene (plural: loci).
Lysogen	A strain of bacteria carrying a prophage; a phage genome integrated into the host chromosome and being replicated passively as part of the replicon that is the host chromosome.
Lysogenic cycle	One of two outcomes of the infection of a host cell by a temperate phage. One outcome is that the phage genome becomes repressed and the phage DNA replicates as part of the host DNA, forming a lysogen; infrequently, the lysogen may become induced and the host cell may burst, releasing a number of phage particles. The other outcome is the lytic cycle that produces progeny phage particles.
Lysogeny	See **lysogenic cycle**.
Lytic cycle	A form of bacteriophage infection cycle in which the phage DNA does not replicate stably with host DNA, but replicates itself and lyses the host cell to release a number of daughter phages.
Macroevolution	Evolution above the species level, leading to the formation of genera, families and other higher taxa; also called **transspecific evolution**.
Map unit	A unit of genetic distance corresponding to a recombination frequency of 1%.
Marker	A gene or allele whose inheritance is under observation in a cross.
Maternal effect	A phenotypic character in the progeny determined by the genotype of the mother, e.g. direction of coiling in snail shells.
Maternal inheritance	Phenotypic character derived solely from the female parent; a pattern of inheritance determined by a genetic determinant carried not in the nuclear chromosomes but in DNA in the cytoplasm or in a cytoplasmic oranelle, e.g. mitochondrion or chloroplast.
Mating type	The mating compatibilities of an organism, which are usually genetically controlled.
Megaspore	The larger of the two kinds of haploid spores produced by vascular plants; the smaller kind is called a **microspore**. In seed plants, the megaspore develops into the embryo

sac (the female gametophyte), while the microspore gives rise to the pollen grain (the male gametophyte).

Meiosis	Two successive divisions of a diploid nucleus following one single replication of the chromosomes, so that the resulting four nuclei are haploid.
Mendelian population	An interbreeding group of organisms sharing a common gene pool.
Merozygote	A partially diploid bacterial cell arising from conjugation, transduction or transformation.
Messenger RNA	An RNA molecule (messenger RNA) whose nucleotide sequence is translated into an amino acid sequence on ribosomes during polypeptide synthesis.
Metacentric chromosome	A chromosome with the centromere near the middle of its length.
Metaphase	The second stage of mitosis or of meiosis (I or II), during which the condensed chromosomes line up on a plane between the two poles of the cell.
Microevolution	Evolution of races or sub-species within a species, or, at most, of new species within a genus.
Microspore	The smaller of two unequal gamete types (cf. **megaspore**).
Mitosis	The division of a nucleus following replication of the chromosomes, so that the resulting daughter nuclei have the same number and types of chromosomes as the parent nucleus.
Modification enzyme	See **restriction enzyme**.
Modifier gene	A gene, alleles of which interact with other genes by modifying their phenotypic expression.
Monoecious	Of organisms (usually plants) having male and female sex organs on the same individual and producing male and female gametes (cf. **dioecious**).
Monosomic	An aneuploid cell, tissue or organism with all except one of a particular chromosome missing, but normal for all other chromosomes.
Monozygotic twins	Twins developed from a single fertilized ovum that gives rise to two embryos at an early developmental stage; also called **identical twins** (cf. **dizygotic twins**).
Mosaic	An individual composed of genetically different groups of cells, each expressing a different phenotype.
mRNA	See **messenger RNA**.
Multiple alleles	The occurrence in a population of more than two different alleles of a gene.
Multiple-factor inheritance	The determination of a phenotypic trait by alleles of more than one gene. See **polygene**.
Mutant	An allele different from a pre-existing one as a result of mutation, or an individual carrying such an allele.
Mutation	The process by which a heritable change occurs in a gene, sometimes used ambiguously to mean the result of the process of mutation (**mutant**).
Mutator gene	A gene at which one or more alleles exist that increase the mutation rate of other genes in the same organism.
Native DNA	Double-stranded DNA isolated from a cell with its hydrogen bonds between strands intact (cf. **denatured DNA**).
Natural selection	The differential reproduction of alternative genotypes due to variable fitness (cf. **artificial selection**).
Non-disjunction	The failure of two sister chromatids or homologous chromosomes to separate during cell division, so that both go to the same pole, thus producing **aneuploid** nuclei.
Non-homologous	Of chromosomes or chromosome segments that contain dissimilar genes and that do not pair during meiosis.

Non-random mating	A mating system in which the frequencies of the various kinds of mating with respect to some trait or traits are different from those expected according to chance.
Nucleolar organizer	One or a few parts of a eukaryotic genome at which the genes for **rRNA** are arranged in long tandem repeats and at which, cytologically, nucleoli are formed.
Nucleolus	A nuclear organelle of eukaryotes, associated with the chromosomal site of the genes coding for rRNA; hence the site of rRNA synthesis.
Nucleoside	A nucleic acid component consisting of a base (purine or pyrimidine) covalently linked to a sugar, either deoxyribose (in DNA) or ribose (in RNA).
Nucleosome	A subunit of chromatin in a eukaryotic chromosome, with a length of DNA duplex wound around a core of histone protein subunits.
Nucleotide	A nucleic acid component consisting of a base (purine or pyrimidine) covalently linked to a sugar, either deoxyribose (in DNA) or ribose (in RNA), and to a phosphate group.
Nucleus	A membrane-enclosed organelle of eukaryotes that contains the chromosomes.
Nullisomic	An aneuploid cell, tissue or organism in which all copies of a particular chromosome are missing.
Ochre codon	One of the three 'stop' codons in the genetic code, UAA.
Oligomer	A protein composed of two or a few homologous polypeptide subunits.
Oncogene	A gene or allele implicated in cancer.
Oocyte	See **oogonium**.
Oogenesis	The process of differentiation of a mature egg cell from an undifferentiated germ-line cell, including the process of meiosis.
Oogonium	A primordial germ cell that gives rise, by mitosis, to **oocytes**, from which the ovum and polar bodies develop by meiosis.
Operator	In the DNA of an operon or regulated gene, a nucleotide sequence that is recognized and bound by a repressor protein, which in turn inhibits transcription of the operon.
Operon	Two or more adjacent structural genes, transcribed into a single mRNA, together with the adjacent transcriptional control sites (promoter and usually operator). This organization places the expression of the structural genes in an operon under the coordinate control of a single set of control sites.
Origin (of replication)	A DNA base sequence at which replication is initiated.
Outbreeding	A mating system in which matings between close relatives do not usually occur.
Ovum	A female gamete.
Pachytene	A stage in meiotic prophase I.
Palindrome	A sequence of symbols that reads identically in both directions, e.g. inverted repeat base sequence in DNA, which in genetics is characteristic of a recognition site or target for a DNA-binding protein, e.g. restriction enzyme or regulatory protein.
Paracentric inversion	A chromosomal inversion that does not include the centromere.
Parental type	An association of genetic markers, found among the progeny of a cross, that is identical to an association of markers present in a parent (cf. **recombinant type**).
Parthenogenesis	The production of an embryo from a female gamete without participation of a male gamete.
Partial dominance	See **incomplete dominance**.
PCR	See **polymerase chain reaction**.
Pedigree	A diagram showing the ancestral relationships among individuals of a family over two or more generations.

123

Penetrance	This term is used of an allele (which may be dominant or recessive), the frequency of whose manifestation in the phenotype is variable. Penetrance is the frequency with which such an allele is manifest.
Peptide bond	The covalent bond formed between the NH_2 group of one amino acid and the COOH group of another, with the elimination of H_2O.
Pericentric inversion	A chromosomal inversion that includes the centromere.
Perithecium	A fruit body, containing a number of asci (singular: ascus), that develops from a **protoperithecium** of an ascomycete fungus.
Petite	A yeast mutant phenotype characterized by slow growth and small colony size. Petite mutants may be nuclear or cytoplasmic.
Phage	See **bacteriophage**.
Phenocopy	A non-hereditary phenotypic modification, produced by environmental causes, that mimics a similar phenotype due to a particular genotype.
Phenotype	The observable characteristics of an individual, resulting from the interaction between the genotype and the environment in which development occurs.
Phenotypic variance	The variance among individuals with respect to some phenotypic trait or traits (cf. **genetic variance** and **variance**).
Physical map	A map of the genome of a species based on physical data, i.e. physical lengths of DNA fragments between restriction sites, onto which gene loci can be positioned based on DNA hybridization data.
Plaque	A clearing in a lawn of bacterial cells created by the growth of a bacteriophage and its concomitant killing of cells.
Plasmid	A genetic element (DNA molecule) harboured within a host cell that replicates independently of the host chromosomes.
Plastid	In plants, a self-replicating organelle that can differentiate into a **chloroplast** or other specialized cytoplasmic structure.
Pleiotropy	An effect whereby a single mutant gene affects two or more apparently otherwise unrelated aspects of the phenotype of an organism.
Point mutation	A mutation caused by the alteration, insertion or deletion of a single base pair in the DNA duplex.
Polar bodies	The smaller cells that are produced during meiosis in oogenesis and that do not develop into functional ova.
Polygene	One of a number of genes each with a minor, additive effect on a phenotypic character.
Polygenic	Of traits determined by alleles of multiple genes, each having only a minor effect on the expression of the trait.
Polymerase	An enzyme that assembles a number of similar or identical subunits into a larger unit, or polymer, e.g. DNA polymerase and RNA polymerase.
Polymerase chain reaction	The *in vitro* technique by which a specific sequence in DNA can be amplified by repeated cycles of differential replication; normally abbreviated to **PCR**.
Polymorphism	The presence of several forms (of a trait or of a gene) in a population.
Polypeptide	A chain of amino acids covalently bound by peptide bonds.
Polyploid	A cell, tissue or organism having three or more complete chromosome sets.
Polyribosome	See **polysome**.
Polysome	A polyribosome consisting of two or more ribosomes bound together by their simultaneous translation of a single mRNA molecule.
Polytene chromosome	An interphase chromosome that has undergone a number of rounds of DNA

replication, without accompanying nuclear divisions, and in which the resulting chromosome strands are paired lengthwise to create a rope-like giant chromosome revealing a specific banding pattern of the chromatin.

Position effect
A change in the phenotypic effect of one or more genes, due to a change in their position within the genome.

Primase
The enzyme that synthesizes the RNA primer normally required for the initiation of DNA replication *in vivo*.

Primer
A short RNA or single-stranded DNA oligonucleotide that will function to initiate DNA replication on a single-stranded DNA template.

Probe
A DNA or RNA fragment used in DNA–DNA or DNA–RNA hybridization assays.

Processing
Post-transcriptional modification of initial RNA transcripts necessary for the maturation of functional mRNA, tRNA or rRNA molecules.

Programmed cell death
Cell death as a part of a normal developmental process, also known as **apoptosis**.

Prokaryote
A cell or organism that lacks a membrane-bound nucleus and does not undergo mitosis or meiosis. The bacteria and blue-green algae are prokaryotes.

Promoter
A nucleotide sequence in DNA at the beginning of a transcription unit that is recognized by RNA polymerase as a site at which to begin transcription.

Prophage
The repressed form of a phage genome present in a lysogen.

Prophase
The first stage of mitosis or of meiosis (I or II) during which the chromosomes condense and become visible as distinct bodies.

Protein
A polymer composed of one or more polypeptide subunits and possessing a characteristic three-dimensional shape imposed by the sequence of its component amino acid residues.

Prototroph
A strain of organisms capable of growth on a defined minimal medium from which they can synthesize all of the more complex biological molecules they require (cf. **auxotroph**).

Race
A population or group of populations distinguishable from other such populations of the same species by the frequencies of alleles of certain genes, chromosomal rearrangements or hereditary phenotypic characteristics. A race that has received a taxonomic name is a subspecies.

Random genetic drift
Variation in allele frequency from one generation to another due to chance fluctuations.

Random mating
A sample of a reproducing population obtained in such a way that each individual in the population, or each allele of any gene in the genome, has the same chance of reproducing.

Recessive
An allele is recessive with respect to another allele if, when together in a heterozygote, the phenotype resembles that of an individual homozygous for the other allele.

Reciprocal crosses
Those in which each of two strains provides the males in one cross and the females in the other, as in male $A \times$ female B and male $B \times$ female A.

Reciprocal translocation
A translocation that involves an exchange of chromosome segments between two non-homologous chromosomes.

Recombinant (type)
An association of genetic markers, found among the progeny of a cross, that is different from any association of markers present in the parents (cf. **parental type**).

Recombination
Processes generating new combinations of genetic material: (1) independent assortment; (2) the exchange of homologous parts between homologous chromosomes, resulting in new combinations of alleles of those genes on the particular chromosome.

Regulatory gene	A gene involved in the control of transcription of a **structural gene**.
Replicon	A self-replicating genetic element possessing a site for the initiation of DNA replication and genes specifying the necessary functions for controlling replication.
Repressor	A protein that binds to an operator sequence in DNA and thereby inhibits the transcription of adjacent genes by blocking RNA polymerase from the promoter for those genes.
Reproductive isolation	The inability to interbreed due to biological differences.
Restriction enzyme	An endonuclease that recognizes specific nucleotide sequences in DNA and then makes a double-strand cleavage of the DNA molecule.
Restriction fragment length polymorphism (RFLP)	A difference in length of a particular section of DNA caused by changes in base sequence which create or remove an adjacent target site for the restriction enzyme being used.
Reversion	A second mutation that reverses, or restores, totally or partially, the genetic information altered by a first mutation.
RFLP	See **restriction fragment length polymorphism**.
Ribonucleic acid	A polynucleotide (RNA) in which the sugar residue is ribose and which has uracil rather than the thymine found in DNA.
Ribosomal RNA	The RNA molecules (rRNA) that are structural parts of ribosomes, i.e. 5S, 16S and 23S RNAs in prokaryotes and 5S, 5.8S, 18S and 28S RNAs in eukaryotes.
Ribosome	An organelle, consisting of two subunits composed of RNA and proteins, that synthesizes polypeptides, whose amino acid sequences are specified by the nucleotide sequences of mRNA molecules.
RNA	See **ribonucleic acid**.
RNA polymerase	The enzyme responsible for transcribing the information encoded in DNA into RNA; also called **transcriptase**.
rRNA	See **ribosomal RNA**.
Selection	See **natural selection** and **artificial selection**.
Self-fertilization	The union of male and female gametes produced by the same individual.
Selfing	Sexual reproduction by self-fertilization.
Sex chromosomes	Chromosomes that are different in the two sexes and that are involved in sex determination (cf. **autosomes**).
Sex-limited	Pertaining to genetically controlled characters that are phenotypically expressed in only one sex.
Sex-linkage	Pattern of segregation of genes that are located in the sex chromosomes.
Sex-ratio	The number of males divided by the number of females (sometimes expressed as a percentage) at fertilization (primary sex ratio), at birth (secondary sex ratio) or at sexual maturity (tertiary sex ratio).
Sickle-cell anaemia	A human disease characterized by defective haemoglobin molecules and due to homozygosity for an allele coding for a defective beta chain of haemoglobin.
Sickle-cell trait	A phenotype recognizable for the sickling of red blood cells exposed to low oxygen tension, and determined by heterozygosity for the allele responsible for sickle-cell anaemia.
Somatic cells	All body cells except the gametes and the cells from which these develop.
Southern blotting	The use of a cloned fragment of DNA to probe for a homologous fragment in a DNA restriction digest.
Speciation	The process of species formation.
Species	Groups of interbreeding natural populations that are reproductively isolated from

other such groups.

Spermatids	The cells that are produced during meiosis in spermatogenesis and that eventually develop into functional spermatozoa.
Spermatocyte	Cell from which spermatids are produced by meiosis.
Spermatogenesis	The process of differentiation of a mature sperm cell from an undifferentiated germ-line cell, including the process of meiosis.
Spermatozoon	In animals, a male gamete.
Spindle	In eukaryotes, an ellipsoidal collection of fibres visible during mitosis and meiosis and involved in the separation of homologous chromosomes or sister chromatids toward opposite poles of the cell.
Sporophyte	The diploid, asexual generation that produces the spores in plants in which a sexual (haploid) generation alternates with an asexual (diploid) generation (cf. **gametophyte**).
Structural gene	A gene that codes for a polypeptide (cf. **regulatory gene**).
Sub-species	A population or group of populations distinguishable from other such populations of the same species by the frequencies of alleles, chromosomal rearrangements or hereditary phenotypic characteristics. Sub-species sometimes exhibit incipient reproductive isolation, although not sufficiently to make them different species.
Supercoil	A double-stranded DNA molecule containing extra twists in the helix that cause the helix to coil upon itself. In order for these extra twists to be maintained, it is necessary that the end of the double-stranded helix be constrained from rotating freely, e.g. in a covalently closed circular DNA molecule.
Sympatric	Of populations or species that inhabit, at least in part, the same geographic region (cf. **allopatric**).
Synapsis	The pairing of chromosomes at meiosis.
Synaptinemal complex	An organelle, present during meiosis, that mediates close pairing between homologous regions of chromatids.
Tandem duplication	A chromosomal duplication in which the duplicated segments are adjacent (in tandem), inverted or not; also called repeat.
Telocentric chromosome	A chromosome with, at least by cytological criteria, a terminal centromere.
Telophase	The fourth and final stage of mitosis, or final stage of meiosis (I or II).
Temperate phage	A bacteriophage capable of lysogenizing a host cell (cf. **virulent phage**).
Template	The DNA single strand, complementary to a nascent RNA or DNA strand, that serves to specify the nucleotide sequence of the nascent strand.
Terminator	A nucleotide sequence in DNA that causes RNA polymerase to cease transcription.
Test-cross	A cross between an individual of unknown genotype (at one or more loci) and the corresponding recessive homozygote.
Tetraploid	A cell, tissue or organism having four chromosome sets.
Tetrasomic	A cell, tissue or organism having one chromosome represented four times.
Topoisomerase	An enzyme that changes the topology of DNA, specifically relating to supercoiling.
Totipotency	The ability of a cell to develop into a complete organism.
Transcription	The transfer of genetic information encoded in the nucleotide sequence of DNA into a nucleotide sequence of an RNA molecule.
Transcription factor	A protein that, together with RNA polymerase, initiates transcription.
Transduction	The transfer of DNA from one cell to another, effected by a virus.
Transfection	In higher organism somatic cell genetics, the incorporation of genetic material from a donor organism into the chromosomes of a recipient cell; synonymous with

	transformation. In bacterial genetics, the infection of a cell with naked viral DNA resulting in the production of infective particles.
Transfer RNA	Special RNA molecules (tRNA) that bind specific amino acids to form aminoacyl-tRNAs and that transfer their amino acids to growing polypeptides during association with ribosomes.
Transformation	(1) The conversion of normal cells into cancerous cells; (2) the direct assimilation of exogenous DNA into a cell by bacteria and microbial eukaryotes (cf. **transfection**).
Transition	A base-pair substitution mutation resulting in the replacement of one purine by another purine, or of one pyrimidine by another pyrimidine (cf. **transversion**).
Translocation	A chromosomal mutation characterized by a change in position of a chromosome segment.
Transposition	A translocation of a chromosome segment from one position to another without a reciprocal exchange (cf. **reciprocal translocation**).
Transposon	A transposable DNA sequence carrying one to many genes bounded at each end by identical (often insertion) sequences, which confer the ability to move from one location to another within DNA.
Transversion	A base-pair substitution mutation resulting in the replacement of a purine by a pyrimidine or vice versa (cf. **transition**).
Triploid	A cell, tissue or organism having three chromosome sets.
Trisomic	A cell, tissue or organism having one chromosome represented three times.
tRNA	See **transfer RNA**.
Turner's syndrome	A human syndrome that is due to monosomy for the X chromosome with absence of a Y chromosome (XO); affected individuals are phenotypically female but usually have underdeveloped gonads.
Univalent	An unpaired chromosome at the first meiotic division.
Variance	A statistical measure of variation around the mean in a population or sample of a population; the average squared deviation of the observations from their mean value.
Vector	An autonomously replicating DNA molecule, derived from a plasmid or virus, into which other pieces of DNA may be cloned.
Vegetative reproduction	See **asexual reproduction**.
Virulent phage	A bacteriophage whose infection invariably kills the host cell rather than lysogenizing it (cf. **temperate phage**).
Wild type	The prevailing phenotypes or the prevailing alleles, if any, in a natural population.
YAC	Yeast artificial chromosome, used as a specialized cloning vector.
Zygote	The diploid cell formed by the union of egg and sperm nuclei within the cell.
Zygotene	A stage in meiotic prophase I.

INDEX

Index

Hfr, *Escherichia coli* cell, 41, 42, 44, 45, 50

homologous recombination, 74

homozygous, 2, 7, 9, 10, 19, 21, 23, 25, 95

human pedigree analysis, 94–106

hypervariable markers, 102

illegitimate recombination, 42, 52, 53, 59

immunity, of lysogen to superinfection, 56

independent assortment, 2, 16, 18, 19, 21, 27, 29, 92

induction, of temperate phage, 40, 58, 60

integration, of temperate phage, 40, 58, 60

integration, of F plasmid, 51

interrupted mating, 42, 44, 45, 49

inversion, chromosomal, 107

lambda, *Escherichia coli* temperate phage, 41, 57, 58, 59

linear tetrad, 3, 5

linkage, 2, 25, 27, 29, 92

linkage:
  group, 29, 31
  in Hfr crosses, 47
  map, 29, 31

locus, 1

lyotonic dystrophy, 102

lysis, as opposed to lysogenization, 56

lysogenic phage, 40, 55

lysogenization, 40, 55, 56, 57

lytic cycle, of phage, 40, 57, 65

map distance, 25, 27, 29, 31

map units, 5

mapping function, 27

maternal inheritance, 14

Mendel's first law, 4, 6, 7, 10, 11, 13

Mendel's second law, 18, 20, 22, 24

Mirabilis, 15

mis-sense mutations/mutants, 86

mitochondrial DNA, 14

mitochondrion, 4

mouse, 10

multiple crossing over, 26

Mus, 10

mutagen, 86

mutant, 86

mutants (mutations):
  conditional, 90
  deletion, 78
  frame shift, 86
  mis-sense, 86
  nonsense, 86
  point, 80
  recessive lethal, 11
  suppresser, 64, 65

mutation, 86

NCO *see* non-crossover

Neurospora crassa, 4, 14, 30, 73, 88

no dominance, 7

non-allelic, 17

non-allelic reversion, 86

non-crossover (NCO), 26, 31, 32, 34, 36, 38

non-parental ditype (NPD), 16, 25, 27, 37

nonsense mutations/mutants, 86

nonsense (stop) codon, 86

non-sister chromatids, 2

NPD *see* non-parental ditype

octad, 4

ordered tetrad, 4, 16

oranelle, 4, 14, 15

organelle DNA, 4, 14, 15

origin of transfer, of Hfr DNA, 44, 50

P22, temperate phage of *Salmonella typhimurium*, 55

packaging, of DNA into phage head, 58

parental ditype (PD), 16, 25, 27, 37

parental (P) generation, 7, 8, 9

partial diploids, from bacterial crosses, 54, 59

PD *see* parental ditype

pedigree analysis, 94–106

phage:
  induction of temperature, 40, 58, 60
  integration of temperate, 40, 58, 60
  lysogenic, 40, 55
  lysogenization by, 40, 55, 56, 57
  lytic cycle, 40, 57, 65
  P22, 55
  T4, 83

temperate, 40, 55

transduction by, 41, 59

virulent, 40, 63

phage head, DNA packaging into, 58

phenotype, 7, 20

Pisum, 19

plaque:
  formed by phage in a lawn of bacteria, 55, 57, 64
  clear, 55, 57
  morphology, 55, 57, 64
  turbid, 55, 57

plasmids, 40, 66

plasmids, conjugative, 41

plasmids, integrating, 41

plasmids, non-conjugative, 41

point mutations/mutants, 80

positive crossover interference, 32, 35, 37

probability, 6, 7, 8, 10, 19, 21, 25

prophage, 40

prophage excision, 40

prototroph, 18

Punnett square, 20

pure-breeding, 7, 9

*r*I mutant plaques, 64

*r*II mutants, 83

random progeny analysis, 5

ratios:
  1:1, 5, 10
  1:1:1:1, 17, 23, 29
  1:2:1, 7, 10
  2:1, 11
  3:1, 10
  9:3:3:1, 22

recessive, 7, 9

recipient cell, in bacterial cross, 41, 44, 61

recombination:
  frequency, 5
  homologous, 25, 74
  illegitimate, 42, 52, 53

replica plating, 61

repressor, encoded by temperate phage, 56, 57

restriction endonuclease, 66

revertant, 86, 87, 88

risk estimation, 99–102